Bioceramics:
Materials and Applications V

Technical Resources

Journal of the American Ceramic Society

www.ceramicjournal.org

With the highest impact factor of any ceramics-specific journal, the *Journal of the American Ceramic Society* is the world's leading source of published research in ceramics and related materials sciences.

Contents include ceramic processing science; electric and dielectic properties; mechanical, thermal and chemical properties; microstructure and phase equilibria; and much more.

Journal of the American Ceramic Society is abstracted/indexed in Chemical Abstracts, Ceramic Abstracts, Cambridge Scientific, ISI's Web of Science, Science Citation Index, Chemistry Citation Index, Materials Science Citation Index, Reaction Citation Index, Current Contents/Physical, Chemical and Earth Sciences, Current Contents/Engineering, Computing and Technology, plus more.

View abstracts of all content from 1997 through the current issue at no charge at www.ceramicjournal.org. Subscribers receive full-text access to online content.

Published monthly in print and online. Annual subscription runs from January through December. ISSN 0002-7820

International Journal of Applied Ceramic Technology

www.ceramics.org/act

Launched in January 2004, *International Journal of Applied Ceramic Technology* is a must read for engineers, scientists,and companies using or exploring the use of engineered ceramics in product and commercial applications.

Led by an editorial board of experts from industry, government and universities, *International Journal of Applied Ceramic Technology* is a peer-reviewed publication that provides the latest information on fuel cells, nanotechnology, ceramic armor, thermal and environmental barrier coatings, functional materials, ceramic matrix composites, biomaterials, and other cutting-edge topics.

Go to www.ceramics.org/act to see the current issue's table of contents listing state-of-the-art coverage of important topics by internationally recognized leaders.

Published quarterly. Annual subscription runs from January through December. ISSN 1546-542X

American Ceramic Society Bulletin

www.ceramicbulletin.org

The *American Ceramic Society Bulletin*, is a must-read publication devoted to current and emerging developments in materials, manufacturing processes, instrumentation, equipment, and systems impacting the global ceramics and glass industries.

The *Bulletin* is written primarily for key specifiers of products and services: researchers, engineers, other technical personnel and corporate managers involved in the research, development and manufacture of ceramic and glass products. Membership in The American Ceramic Society includes a subscription to the *Bulletin*, including online access.

Published monthly in print and online, the December issue includes the annual *ceramicSOURCE* company directory and buyer's guide. ISSN 0002-7812

Ceramic Engineering and Science Proceedings (CESP)

www.ceramics.org/cesp

Practical and effective solutions for manufacturing and processing issues are offered by industry experts. CESP includes five issues per year: Glass Problems, Whitewares & Materials, Advanced Ceramics and Composites, Porcelain Enamel. Annual subscription runs from January to December. ISSN 0196-6219

ACerS-NIST Phase Equilibria Diagrams CD-ROM Database Version 3.0

www.ceramics.org/phasecd

The ACerS-NIST Phase Equilibria Diagrams CD-ROM Database Version 3.0 contains more than 19,000 diagrams previously published in 20 phase volumes produced as part of the ACerS-NIST Phase Equilibria Diagrams Program: Volumes I through XIII; Annuals 91, 92 and 93; High Tc Superconductors I & II; Zirconium & Zirconia Systems; and Electronic Ceramics I. The CD-ROM includes full commentaries and interactive capabilities.

Bioceramics: Materials and Applications V

Ceramic Transactions Volume 164

*Proceedings of the 106th Annual Meeting
of The American Ceramic Society,
Indianapolis, Indiana, USA (2004)*

Editors
Veeraraghavan (V) Sundar
Richard P. Rusin
Claire A. Rutiser

Published by
The American Ceramic Society
PO Box 6136
Westerville, Ohio 43086-6136
www.ceramics.org

For information on ordering titles published by The American Ceramic Society, or to request a publications catalog, please call 614-794-5890, or visit our website at www.ceramics.org

ISBN 1-57498-185-4

Contents

Preface

It is once again our privilege and pleasure to edit these proceedings of the Bioceramics symposium held during the 2004 Annual Meeting of The American Ceramic Society.

The last year has been outstanding, in the apparent and sudden recognition by the mass media, at least in the US, of the increasing trend in body modification by the public at large. Leaving aside the social and cultural implications of this trend, we have enjoyed an unusual recognition of the tools and materials the healing professions use to enhance our lives. We do indeed live in interesting times. As an aside, at least two of the editors (Sundar and Rutiser) wish it had not been "reality television" spectacles that triggered this phenomenon. Also, at least two of the editors (Sundar and Rusin) are somewhat puzzled by some people's insistence that extensive and invasive elective cosmetic surgery is a dream come true, but a much-needed trip to the dentist is something to be dreaded.

In such times, the range of topics presented at the Bioceramics symposium has been, to say the least, diverse. The research presented ranged from carbon composite electrodes for neuronal activity sensing, through the influence of microbiological processing on chemical, mineralogical and granulometrical structures of clay loams of Kyrghyzstan, to topics that seem mundane by comparison—such as bioactive zirconia scaffolds with hydroxyapatite and glass coatings.

We hope that these proceedings reflect our intent in highlighting the variety of research topics presented. A specific concentration on interfaces, so critical in integrating biological and inorganic or organic materials, was an area of focus. Apatites and active bone substitute materials are well represented, as always. A small section on dental ceramics is also included here. We hope you enjoy them, and find them as educational as we did.

In closing, we would like to recognize our fellow symposium organizers Alexis Clare, Gary Fischman, Vic Janas, Irene Peterson, Subrata Saha, and Warren Wolf for their enthusiasm and welcomed help, in organizing and conducting this symposium. We are most grateful to the staff of The American Ceramic Society, our sponsoring Divisions (Engineering Ceramics, NICE, Glass & Optical Materials), student pages, and volunteer session chairs, for the hard work that makes these meetings possible. Our colleagues at Dentsply Prosthetics, 3M ESPE and CRMA have generously lent us their expertise and guidance. And once again, to our families, for their support and understanding while we juggled these proceedings with our other commitments, our love and thanks.

Veeraraghavan (V) Sundar
Richard P. Rusin
Claire A. Rutiser

Bioceramic – Metal Interfaces

INTERFACE DIFFUSION/REACTION IN THE HYDROXYLAPATITE COATED CoCrMo ALLOY

Celalettin Ergun, Zafer Evis*,#, Robert H. Doremus*
Mechanical Engineering, Istanbul Technical University, Taksim, Istanbul, TURKEY;
*Materials Science & Engineering, Rensselaer Polytechnic Institute, Troy, NY, 12180
#Presenting Author (evisz@rpi.edu)

ABSTRACT

The purpose of this study is to develop a better understanding of the interface reaction between hydroxylapatite and CoCrMo alloys. Hydroxylapatite powder was sintered onto rods of the alloys by hot isostatic pressing at 900, 1000 and 1100°C at a pressure 120MPa. The reaction at the interface between the components was monitored with EPMA line scanning technique. EPMA analysis on the interface between hydroxylapatite and CoCrMo alloy showed interdiffusion profiles of all the hydroxylapatite and alloy elements and Cr accumulation at the interface. This finding suggests that there was an interfacial bonding provided by diffusion and reaction.

INTRODUCTION

Hydroxylapatite (HA) is generally used as a coating material on metals such as Cobalt-Chromium-Molybdenum (CoCrMo) alloys, stainless steels and titanium alloys, in many load-bearing biomedical applications to take advantage of both the biocompatibility of HA and the strength of the metallic substrates. However, the interface between HA and metals has been a matter of concern for the long-term reliability of the implants in service from the mechanical properties standpoint. In order for the HA coating to be effective and reliable, the coating must be strongly bonded to the metallic surface. For these reason, chemical bonding derived from the reaction/diffusion at the HA-metal interface could develop a highly reliable interface in response to this requirement [1-3]. The presence of reaction/diffusion bonding at the interface may increase the bond strength between HA and the metal, however, it may also cause toxic reactions in the body, since this new interface material has a different composition from HA or the metal [4]. Thus the knowledge of the interface phenomena including the reaction kinetics and the properties of byproducts will certainly help to evaluate the overall interface behavior leading to the design of better implants.

CoCrMo alloys are widely used metallic biomaterials in implants. One of the advantages of these alloys as a substrate material in the coating applications is their close coefficients of thermal expansion to HA compared to titanium alloys decrease the risk of failure in the HA coatings both in the processing and the service.

The aim of the present study is to examine the bonding of HA to CoCrMo alloy and to develop a better understanding of the interface reaction between HA and CoCrMo.

EXPERIMENTAL PROCEDURE

HA coated CoCrMo were prepared with hot isostatic pressing inside evacuated borosilicate glass tubes. The specimens were used to examine the interface between HA and CoCrMo alloy in terms of diffusion and reaction. The sample preparations and experimental methods are discussed in the following:

HA was produced with a precipitation method. Details on this technique are reported in previous papers [5,6]. After the synthesis of HA, the bulk were first crushed with a mortar and pestle to below a 3mm particle size, and then further ground to below a 200-mesh size.

HA powder was calcined at 1100°C for one hour. Then the powder was coated on the alloy rods with a cold press die. The rods were made of CoCrMo alloys and were 25mm in length and 5 mm in diameter. The coating pressure was 200MPa.

The HA coated alloy samples and cold-pressed composite powder were coated with boron nitride aerosol spray and dried at 60°C overnight. The rod samples were placed into borosilicate glass tubes. Boron nitride powder, which had been stored in the furnace at 120°, filled the gap between the samples and the glass tube wall. Next the tube was evacuated and sealed with an oxyacetylene torch and placed into a graphite supporter. The supporter was placed into the HIP chamber. Then the chamber was evacuated to 100-militorr pressure and heated to 750°C. At this temperature, the chamber pressure was increased to 120MPa over 30 minutes with a gas compressor. The samples were hipped at 850°C, 1000°C and 1100°C as final processing temperatures for 2 hour at 120MPa pressure. The heating and cooling rates were 600-degree/hr.

The internal metal-ceramic interfaces of the HA-coated samples were exposed by sectioning and analyzed with the electron microprobe, JEOL Superprobe 733, for interface diffusion of the atomic species across the interface. Analyses were performed at a 15kV acceleration voltage with a focused beam with a diameter of 1µm. A line scan analysis was run for each sample perpendicular to the interface, starting from 15µm inside ceramic. Each step was chosen as 1µm for the best resolution in this instrument.

RESULTS and DISCUSSION

Elemental analyses were achieved at the interface between HA and metal substrate. A SEM picture of a typical interface and scan is shown in Fig. 1.

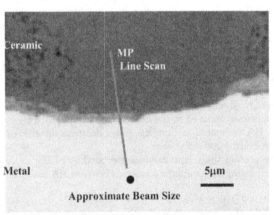

Figure 1: A typical interface where Electron Microprobe line scan analyses were performed.

Elemental analysis was performed for the metal atoms: Co, Cr, and Mo and the HA atoms: Ca, P, and O. At this temperature level, diffusional spreading of the elements was observed. Diffusion profiles of HA coated CoCrMo, hipped at 850°C, 1000°C, and 1100°C are presented in Fig.2, Fig.3, and Fig.4, respectively. The interface diffusion profiles in these samples also showed typical diffusion behavior. Total diffusion zone thicknesses in the samples hipped at 850°C, 1000°C, and 1100°C were about 5μm, 6.5μm 8μm, respectively.

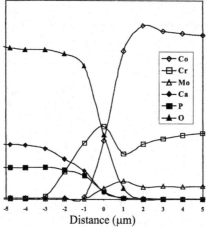

Distance (μm)

Figure 2: Diffusion profiles at the HA/CoCrMo interface after heating in hot isostatic pressing at 850°C for 2 hours at 120MPa pressure.

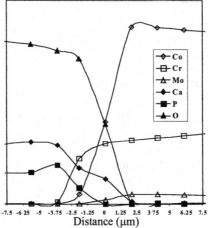

Distance (μm)

Figure 3: Diffusion profiles at the HA/CoCrMo interface after heating in hot isostatic pressing at 1000°C for 2 hours at 120MPa pressure.

HA and Co-Cr-Mo alloy bonding is a complicated interdiffusion process with several atomic constituents. Interdiffusion coefficients (D) can be estimated by the formula:

$$x = 2\sqrt{Dt} \qquad (1)$$

in which, x is the mean diffusion distance in time (t). In Table 1, mean diffusion distances derived from Figures 1 and 2 are listed as well as D values calculated from equation 1. The calculated values are given in Table 1. The D value of $1.1 \times (10)^{-11}$ cm^2/s after 1100°C sintering can be compared with a similar value for interdiffusion of HA and titanium of $4 \times (10)^{-11}$ cm^2/s at 1100°C [4].

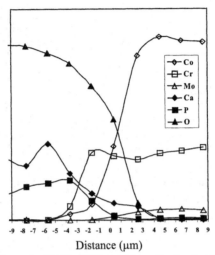

Distance (μm)

Figure 4: Diffusion profiles at the HA/CoCrMo interface after heating in hot isostatic pressing at 1100°C for 2 hours at 120MPa pressure.

Table 1: Approximate diffusion zone thickness, mean diffusion distance and apparent diffusion coefficients at different processing conditions.

Approximate Interdiffusion Coefficients for HA on CoCrMo Alloy			
Temperature (°C)	Diffusion Zone Thickness (μm)	Mean Diffusion Distance (μm)	$Dx10^{12}$ (cm^2/s)
850	5	2.5	4
1000	6.5	3.2	7
1100	8	4	11

Cr seemed to be the deepest penetrating metallic agent into the HA side; it also accumulated at the metal side near the diffusion zone with a typical diffusion profile (Figure 5). This suggested the existence of an interfacial reaction at the interface zone.

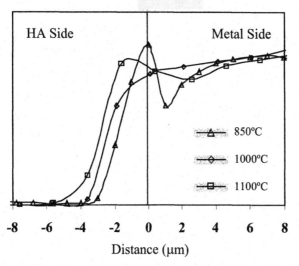

HA Side   Metal Side

                                 ▵ 850°C

                                 ◇ 1000°C

                                 ▢ 1100°C

-8   -6   -4   -2   0   2   4   6   8

Distance (μm)

Figure 5: Diffusion profiles of Cr at the HA/CoCrMo interface at different hipping temperatures.

CONCLUSION

These experimental results showed that HA and the CoCrMo alloys reacted and interdiffused at high temperatures to form a region of continuous chemical bonding. This bonding should give a good strength at the interface region between HA and these alloys. These alloys are especially favorable for HA coatings because their coefficients of thermal expansion are close to that of HA. The question remains whether the interface reaction products are biocompatible.

REFERENCES

1. K. DeGroot, R. Geesink, C.P.A.T. Klein, P. Serekian, "Plasma sprayed coating of hydroxyapatite," *Journal of Biomedical Materials Research*, **21** 1375-81 (1987).

2. J.F. Kay, "Bioactive surface coatings: Cause for encouragement and caution," *JOI,* **14** [1] 43-54 (1988).

3. R.H. Doremus, "Bioceramics," *Journal of Materials Science,* **27** [2] 285-97 (1992).

4. C. Ergun, R.H. Doremus, W.A. Lanford, "Hydroxylapatite-Titanium Interfacial Reactions," *Journal of Biomedical Materials Research,* **65A** 336-43 (2003).

5. C. Ergun, T.J. Webster, R.H. Doremus, R. Bizios, "Hydroxylapatite with substituted Mg, Zn, Cd, and Y: I Structure and Microstructure," *Journal Biomedical Materials Research,* **59** [2] 305-11 (2002).

6. C. Ergun and R.H. Doremus, "Thermal Stability of the Hydroxylapatite-Titanium and Hydroxylapatite-Titania Composites," *Turkish Journal of Environmental Science and Engineering,* **27** 423-29 (2003).

STRUCTURAL AND CHEMICAL CHANGES TO PLASMA SPRAYED HYDROXYAPATITE COATINGS IN SIMULATED BODY FLUID.

Susan Essien Etok & Keith Rogers
Faculty of Medicine & Biosciences
Cranfield University, Shrivenham
Wiltshire. SN6 8AL. UK

ABSTRACT

This prospective study assesses the biological behaviour (bioactivity) of calcium phosphate coatings in a physiological environment. The structural and chemical changes occurring in plasma sprayed hydroxyapatite coatings were investigated by soaking the coatings in simulated body fluid (SBF) for a period of up to 3 weeks. Microstructural changes to coatings were analysed by X-ray diffraction (XRD) using Rietveld analysis with a fundamental parameters approach. Fourier Transform infra-red spectroscopy (FTIR) and scanning electron microscopy (SEM) were used for further chemical and morphological examination.

Our results show that with immersion time the amorphous calcium phosphate (ACP) preferentially dissolves followed by the precipitation of carbonated apatite. There is a increase in diffraction peak broadening and hence crystallite size and microstrain. Due to the increase of crystal anisotrophy with immersion time, it is not possible to separate the components by the Williamson-Hall method. Rietveld analysis has highlighted subtle changes in the lattice parameters of the coatings. The contraction of the c-axis and increase in the a-axis confirms that inclusion of foreign ions has occurred in the crystal lattice.

INTRODUCTION

Calcium hydroxyapatite (HAP: $Ca_{10} (PO_4)_6 (OH)$) coatings on metallic prosthesis have been used extensively in dental and orthopaedic surgery due to their biocompatibility and osteoconductive properties [1-4]. Clinical trials [5] have shown that bioactive materials such as HAP are capable of interacting with the surrounding bone and are thought to produce direct attachment of the implant to bone without an interposed fibrous tissue layer because of the presence of free calcium and phosphate ions at the surface. The fundamental mechanisms that form the basis for the use of hydroxyapatite in implantology is described elsewhere [1].

Plasma spraying (PS) is the current majority manufacturing technology for production of ceramic coatings on metallic substrates. [4,8,9]. Consequently, previous research has revealed that this high temperature fabrication and rapid cooling process imparts some undesirable microstructural and structural transformations due to partial melting and rapid solidification of feedstock particles.[8, 10-13] PS coatings typically contain crystalline phases with an amorphous phase [13] and various metastable phases, each having different solubility's e.g.: tetracalcium phosphate ($Ca_4 (PO_4)_2O_2$), α-tricalcium phosphate (α-$Ca_3 (PO_4)_2$), β-tricalcium phosphate (β-$Ca_3 (PO_4)_2$), and calcium oxide [14, 15].

The presence of amorphous and metastable phases has raised concerns about the long-term robustness of P.S HAP coatings [18].

A fundamental parameter to be considered when HAP coatings are to be used for bone regeneration purposes is material bioactivity [19, 20]. This parameter is related to the material capability of inducing the precipitation of a new carbonated apatite "bone-like" apatite, on the ceramic surface. The precipitation of the bone-like apatite is said to depend on the HAP dissolution .

Of all coating characteristics, partial dissolution of HAP coatings is mandatory to catalyse new bone growth. However, extremely rapid dissolution of amorphous and metastable phases leads to poor bone bonding and poor coating adhesion and cohesion.[4, 13, 21] Despite the rapid dissolution, coatings with a high percentage of amorphous phase present have been shown to be beneficial to inhibit crack formation and propagation within the coating whilst accelerating bone fixation. [22] In contrast *Tong et al., (1998)* have recommended that the optimal phase contents of HAP coatings be principally composed of crystalline HAP and nanocrystals of HAP giving rise to a broad peak between 32.0° 2θ and 32.4° 2θ on a XRD pattern (for Cu Kα radiation)[13]. *Khor et al., (2004)* have suggested that HAP coatings with a small amorphous content (>15%) would be preferred for accelerated fixation of implants. [14] In light of contradictory reports, the authors postulate that a prerequisite to the optimization of the phase composition of as-received coatings (and hence osteotropism) is an in-depth understanding of the possible *in vitro* changes to the bone-bonding surface of the coatings.

In contrast to previous studies, which investigated only the relationship between coating characteristics and dissolution [19, 23-25], the objective of this study was to highlight the microstructural (in terms of crystallite sizes and microstrain) and chemical changes (and hence bioactivity) occurring in P.S coatings in a body-analogous environment. It has been proposed previously [4],that structural and chemical properties of coatings will change after different degrees of dissolution, which will ultimately affect clinical performance. This study is critical to improve the understanding of the dissolution-precipitation mechanisms of P.S coatings and aid further improvements to the osteotropism of the coatings.

METHODS AND MATERIALS

Coating fabrication

The coatings were deposited on titanium substrates (test coupons, diameter~2cm), that were initially cleaned in ethanol in an ultrasonic bath, rinsed in deionised distilled water, and dried in a stream of air. Conventional plasma spraying methods complying to BS ISO 13779-2:200 (Implants for surgery. Hydroxyapatite. Coatings of hydroxyapatite) were used.

Dissolution studies

Previous studies [26] revealed that the ion concentration of SBF closely resembles the concentration of human blood. SBF was prepared as described by *Kokubo et al., (1994)*. The coated specimens were immersed in either 100ml of SBF or water for periods of 1, 3, 7, 14, and 21 days before being removed, washed in deionised water and air-dried pending further analysis. All experiments were performed under aseptic conditions in a sealed beaker at 37°C in an incubator.

Diffraction analysis
    The coatings were analysed *in situ* by conventional powder X-ray diffractometry. Diffraction data was collected by means of a Philips PW1830 diffractometer fitted with a diffracted beam monochromator to produce diffractograms from CuKα wavelengths. Phase identification was achieved with reference to the International Centre for diffraction data database using the software 'CSM' (Oxford Cryosystems).
    A standardless Rietveld analysis as described by *Rogers et al., (2004)* was undertaken to obtain a quantitative phase analysis of the coatings in question and accurate lattice parameters of crystalline phases [27]. The analysis was undertaken using TOPAS (Bruker-AXS) and structural models acquired from the Inorganic Crystal Structural Database. A fundamental parameters approach (FPA) was used whereby instrument and sample contributions are calculated from first principles by the software. Measurable geometric parameters are used in order to generate profile shapes. In addition, Bragg angles are corrected and profile shapes accurately described. The convultion process used in this FPA can be represented as:

$$Y(2\theta) = (W \times G_{eq} \times G_{ax}) \times S \times P \times U + Bkg \qquad (1)$$

Where:
$Y(2\theta)$ = final observed profile
W = emission profile
G_{eq} = equatorial instrument contributions
G_{ax} = axial instrument contributions
S = various sample contributions
P = phases dependent microstructural effects
U = user-defined convultions
Bkg = background

In contrast to conventional direct convultion approaches, there is no need for an instrument profile function to be determined by the measurement of strain-free standards with an infinite crystallite size. Cheary, R.W.& Coelho, A. A. (1988a).

Microstructural analysis was performed using a Pseudo-Voigt analytical profile ($K\alpha_1$ & $K\alpha_2$) as described elsewhere. Independent estimates of Integral breadth (IB) and peak positions were obtained from this fitting procedure. Peak broadening attributable to the specimen microstructure was estimated using:

$$IB_f = (IB_h^2 - IB_g^2)^{\frac{1}{2}} \qquad (2)$$

where subscripts f, h & g refer to the structural, observed and instrument profiles respectively[27]. The instrumental broadening component was determined from diffraction data obtained from a NIST silicon standard reference material,

NBS640b. Williamson-hall analysis was used for the assessment of the crystallite size and microstrain contributions to the peak broadening as described previously using the Scherrer equation [28, 29].

SEM analysis
Scanning electron microscopy was used to observe the changes in coating morphology before and after each immersion period.

FTIR analysis
X-ray diffraction studies were complimented by a Fourier Transform Infrared spectroscopy (VECTOR FT-IR/NIR Spectrometer with OPUS 3.1 software) study. The HAP coatings were analysed in the $4000cm^{-1}$ – $400cm^{-1}$ region, before and after each period of immersion. The potassium bromide discs used for the analysis were prepared with powder scrapped from the coating. The spectra obtained were analysed individually using peak-fit software.

RESULTS
X-ray diffraction

X-ray diffraction analysis of the test samples (See figure 1) revealed peaks matching calcite had been formed after the first day of immersion. The percentage of calcite formed then decreased and disappeared by the fourteenth day of immersion. No phases other than crystalline apatite and amorphous calcium phosphate were identified in the test and control samples. The absence of extraneous X-ray diffraction peaks, demonstrated that no apparent change in the crystallographic phase had occurred upon immersion [30].

An increase was observed in the HAP:ACP ratio for both the test and control samples. A more rapid increase of ACP /HAP was observed in the control samples than in the test samples. On average the increase of the HAP:ACP ratio was approximately 9% and 33% for the test and control samples respectively.

Rietveld refinement of the data from samples immersed in SBF showed an increase in the a-axis dimension and a decrease in the c-axis dimension with dissolution time. In contrast there was no significant change in a-axis dimensions for the control samples with dissolution time. However, the c-axis dimension of the control samples showed similar behaviour to the samples immersed in SBF. The change in unit cell dimensions of the test samples was significantly different from the control samples. The incremental expansion in the a-axis dimensions for the control samples can be explained by the incorporation of H_2O into the apatite crystal structure.[31]

The peak widths (002)-(004) (i.e c-axis direction) showed an increase with immersion time. This increase in peak width is due to a decrease in crystallite size or strain. This possible decrease in crystallite size may be due to small crystallites being precipitated on the surface of the coatings, a decrease in crystallite size of the material present or a combination of the two. The construction of a Williamson hall plot for both the control and SBF sample showed an increase of anisotropy in the crystals. The anisotropy observed was less in the control samples compared to the test sample. Due to the aniostropy, it the bulk changes in crystallite size and strain within the lattice remains unclear. There was an observed increase in strain corresponding to (002) and (004) reflections with immersion time in the test samples and to a lesser extent in the control samples. However, it is not possible to give an absolute value for the bulk crystallite size and microstrain contribution because of the scatter amongst the points in the plot. See figure 1.

Figure 1 (a) Williamson-hall plot of samples immersed in SBF (b) Simple XRD pattern for samples immersed in SBF
See figure 1.

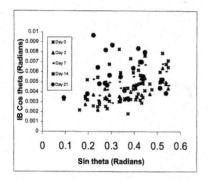 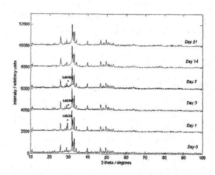

Scanning Electron Microscopy

The surface morphologies of the HAP coatings after soaking in SBF solution for 1,3,7,14 and 21 days are shown in Figure. The as-received coatings had the typical morphology of plasma-sprayed coatings. This consisted of a lamellar microstructure composed of well-flattened splats and spherical droplets with tiny pores and micro-cracks [32, 33]. Essentially, plasma spraying is a rapid heating and rapid cooling and solidification process, which, typically results in the formation of crystalline and metastable deposits of distinct morphology [33]. After the first day of immersion in SBF, it was apparent that the spherical droplets and flattened splats on the surface of the coatings had undergone some dissolution. The surfaces of the coatings show signs of increased roughness and micro-crack propagation. At higher magnification (5000 X), it is observed that the coating is covered in a newly formed layer of small granular structures. This dune-like film is characterized by numerous cracks of tortoiseshell character, which spreads along the whole surface of the ceramic coating. As the immersion duration increased, the dune-like layer became denser and the granules in the layer became larger and hence more visible by day 7 at lower magnification (500X). Other authors have reported comparable results in dissolution studies [34-38].

SEM analysis of the control samples revealed that although dissolution comparable to the samples immersed in SBF had occurred, significantly fewer newly formed precipitates appeared on the surface of the material. By the third week of immersion, the clustered spherulitic precipitates observed in the test samples were not visible at low magnification. *Weng et al., (1997)* have used the classical nucleation theory [39] to explain this phenomenon by the fact that the nucleation and growth of the apatite layer that is the result of the increased ion concentrations by dissolution will occur only when the favourable supersaturation is reached for heterogeneous nucleation or when the interfacial energy has been lowered [34].

Figure 2. SEM images of Surface morphologies of HAP samples before and after immersion in (a) SBF solution and (b) Water

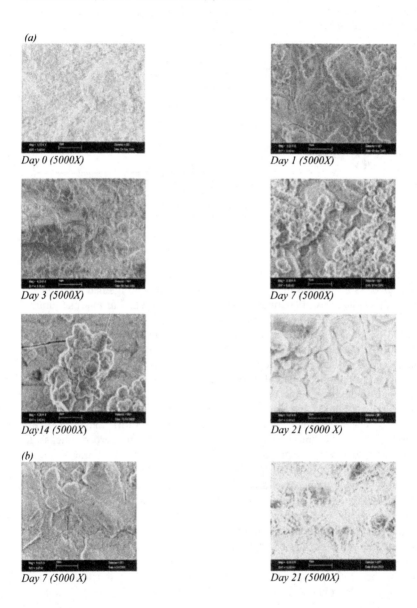

(a)

Day 0 (5000X)

Day 1 (5000X)

Day 3 (5000X)

Day 7 (5000X)

Day14 (5000X)

Day 21 (5000 X)

(b)

Day 7 (5000 X)

Day 21 (5000X)

FTIR

IR vibrational bands peaking at 3300-3400cm^{-1} were detected in both the control and test samples after immersion. These broad bands are indicative of H_2O adsorbed in the materials. The intensity of the peaks in this region increased with immersion time, thus indicating contamination due to water. See figures 3 (a) and (b).

A sharp but weak hydroxyl stretch bond vibration is observed at approximately 3570 cm^{-1} in all as-received samples. With increasing immersion time in SBF, the peak decreases and eventually disappears by day 14. This suggests that there is a decrease in number of hydroxyl ions in the samples. In contrast, the hydroxyl stretch bond vibration is visible in the control samples, although it is reduced in comparison to the as-received samples.

No peaks corresponding to carbonate were detected in the control samples until day 14. In contrast, in the test samples, the appearance of CO_3 vibrational bands at approximately 1427cm^{-1}, 1450cm^{-1} and 870 cm^{-1} was observed after the first week of immersion. This indicates that carbonate-substituted calcium phosphate had been precipitated during immersion. This is believed because the phosphate bands remained present. This is consistent with results obtained by *Queiroz et al., (2003)*. As the calcium phosphate layer formed in the test samples after the seventh day of immersion, it can be concluded that its origin is the $NaHCO_3$ contained in the SBF [40].

Figure 3. FTIR spectra showing a change in chemistry of HAP coating with dissolution time when immersed in (a) SBF and (b) water.

(a)   (b)

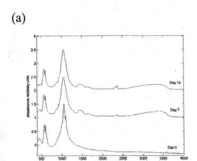 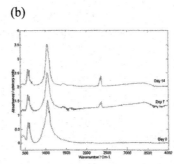

Broad IR bands from 3700- 2500 cm^{-1} and 1650 cm^{-1} were detected in both the test and control samples after immersion. These broad bands indicate that there has been some water adsorption into the materials [30, 34]. Despite the fact that this technique is not strictly quantitative [30], it appears that there is a linear relationship between immersion time and peak intensity due to water adsorption.

Two IR bands peaking at approximately 2360 and 2340 cm^{-1} were detected in the as-received and immersed samples, increasing in intensity with immersion time. This is consistent with results obtained by *Panda et al., 2001*. These peaks have been previously assigned to soluble carbon dioxide [41].

Phosphate ions have four infrared vibrational modes ($\gamma_1 - \gamma_4$), which are have been observed to be active in an apatite lattice [30, 34]. The two strong bands found at approximately 570 and 604 cm^{-1} and the weak shoulder at approximately 550 cm^{-1} have been previously [42] attributed to the three factor group components of the triply de-generate asymmetric O-P-O bending mode (γ_4). Weak bands appeared at approximately 470 and 435 cm^{-1}, which can be assigned to the doubly degenerate γ_2 O-P-O bending mode. This assignment is consistent with previous work. [42-44]. According to *Park et al., (1998)*, the non-degenerate symmetric stretching mode (γ_1) usually appears at approximately 960 cm^{-1}. However, no distinct vibration was detected at this location. Overlapping γ_1 and γ_3 phosphate bands are also detected a broad peak in the wave-number interval of 1300 - 900 cm^{-1} region. Figure 3.3 shows the changes in spectral characteristics of the ceramic materials with immersion time.

DISCUSSION AND CONCLUSIONS

When immersed in SBF solution, both dissolution and precipitation of the ceramic coatings occurred. It appeared that initially, there was significant dissolution of the amorphous phase of the materials. This can be concluded by the reduction in the percentage of amorphous material shown by Rietveld analysis. Crack propagation, increased porosity and surface roughness on the surface of the ceramic materials were a result of the diffusion of ions from the coating surface and the SBF solution. Others [16, 32, 33], have reported the same phenomenon in previous dissolution studies. It is believed that the surface roughness of the ceramic coatings due to dissolution of the amorphous component provided nucleation sites for apatitic precipitation. [32, 33] *Weng et al., (1997)* [32] claim that the nucleation sites were produced due to the existence of defects in the surface molecular layer to forms the steps of nucleation and hence lower the interface energy. It is these two resultant effects of dissolution that have been attributed to provide favourable microenvironment for heterogeneous nucleation, while the roughened surface provides nucleation sites with lower interfacial energy for apatitic precipitation to anchor. [32] Our results showed that the as-received coatings liberate Ca^{2+} and PO_4^{3+} by the dissolution of the amorphous component of the coating. This in turn resulted in localized supersaturation in the microenvironment.

This initial dissolution appeared to be followed by a cohesive apatite crystal precipitation. By virtue of the subtle changes in the unit cell dimensions, microstrain and FTIR data, it is possible to conclude the precipitation was carbonated apatite. In accordance with the Classical nucleation theory [39], at the point of localized supersaturation of Ca^{2+} and PO_4^{3+} in media, these ions migrated from the SBF solution to the surface of the coating. As a result of the high degree of localized supersaturation, apatite nuclei are rapidly formed on the surface are the origin of precipitation of the carbonated-apatite. [45, 46] These findings are consistent with all data generated in this study. EDX results indicate the presence of trace contaminant ions such as Mg^{2+} and Cl^{2-} ions on the surface of the ceramic coatings.

These contaminant ions may have incorporated into the apatite structure, but due to the potential small quantity produced, we have not been able to detect this.

· The nucleation and growth of carbonated or "bone-like" apatite in biphasic apatite coatings is closely associated with the amorphous phase, which is metastable with higher internal energy and hence the amorphous component dominates the dissolution process. [32] However, according to *Ding et al., (2001),* dissolution and precipitation occurs simultaneously in apatite coatings, although at the early stage, dissolution is the more predominant mechanism. Furthermore, it is hypothesized that when the precipitation process catches its speed an apatite layer forms that effectively seals the underneath surface, dissolution stops. Our results confirm this hypothesis, because the rate of ACP dissolution in the later stages of dissolution in SBF solution (day 14-day 21) is significantly less that in the initial stages (day 0-day 7).

There was a marked difference in dissolution behaviour of the samples immersed in SBF solution and the control samples. Significantly less, apatite precipitate had formed on the surface of the control samples. *Weng et al., (1997)* have used the classical nucleation theory [39] to explained this phenomenon by the fact that the nucleation and growth of the apatite layer that is the result of the increased ion concentrations by dissolution will occur only when the favourable supersaturation is reached for heterogeneous nucleation or when the interfacial energy has been lowered. [32] SBF solution contains calcium and phosphate ions and once dissolution of the ceramic coatings commences, supersaturation is reached more rapidly than in the control media which does not contain as high a concentration of free calcium and phosphate ions. *Paschalis et al., (1995)* [47] have postulated that regardless of the crystalline phase, HAP will continue to dissolve as it is subjected to an under-saturated environment. This corroborates our observations that the control samples experienced a 33% loss in ACP compared to a 9% loss of ACP in the test samples over the 21-day immersion period.

XRD and FTIR analysis revealed that there was a significant difference in the nature of apatite formed on the surface of the control and test samples. The apatite precipitation on the samples immersed SBF solution was carbonated apatite whereas for the control samples, was essentially non-carbonated apatite. No significant change was observed in the a-axis unit cell dimensions of the control samples, thus indicating that little or no heterogeneous substitution had occurred in the apatite structure. FTIR analysis revealed that a weak and broad band had appeared at approximately $1370\text{-}1550 \text{cm}^{-1}$. This band has previously been assigned to carbonate. Due to the fact that de-ionised distilled water was used in this investigation, it is probable that this carbonate had originated from small quantities of dissolved CO_2 in the water.

Despite the superficial understanding of the exact mechanisms involved, *in vivo* coating resorption can be categorised as two simultaneous processes; namely dissolution-precipitation at neutral pH (between bone and the coating through ionic species, Ca^{2+}, OH^-, PO_4^{3-}. carried by body fluid) and osteoclastic bone modelling. [1, 3, 4, 19] Nonetheless, dissolution-precipitation remains the key factor affecting bioactivity, which is inherently affected by characteristics of both the HAP coatings and the physiological environment [3].

REFERENCES

1. Sergo, V., Sbaizero, and D. Clarke, *Mechanical and chemical consequences of residual stresses in plasma sprayed hydroxyapatite coatings.* Biomaterials,1997. **18**: p. 477-482.

2. MacDonald, D.E., Betts, F., Stranick, M., Doty, S., Boskey, A.L., *Physicochemical study of plasma-sprayed hydroxyapatite-coated implants in humans.* 2000.

3. Sun, L., et al., *Surface characteristics and dissolution behaviour of plasma-sprayed hydroxyapatite coating.* Journal of Biomed Mater Res, 2002. **62**: p. 228-236.

4. Sun, L.B., C.C; Gross, K.A; Kucuk A., *Material Fundamentals and Clinical Performance of Plasma sprayed Hydroxyapatite Coatings: A Review.* 2001, Centre for Thermal Spray Research, University of New York at Stony Brook: New York. p. 570-592.

5. Le Geros, J.P., et al., *X-ray diffraction method for the quantitative characterisation of calcium phosphate coatings,* ed. E. Horowitz and E. Parr. 1993.

6. Tampieri, A., Celotti, G., Sprio, S., Delcogliano., A., Franzese.,, *Porosity-graded hydroxyapatite ceramics to replace natural bone.* Biomaterials, 2001(22): p. 1365-1370.

7. Toth, J.M., K.L. Lynch, and D.A. Hackbarth, *Ceramic-induced osteogenesis following subcutaneous implantation of calcium phosphates.,* in *Bioceramics,* P. Ducheyne and D. Christiansen, Editors. 1993, Butterworth-Heinemann: Philadelphia,. p. 9-14.

8. Lee, I., Whang, C., Kim, H., Park, J., Song, J.H., Kim, S., *Various Ca/P ratios of thin calcium phosphate films.* Materials Science & Engineering, 2002(C22): p. 15-20.

9. Pilliar, R.M., Filiaggi, M.J.,,, *New calcium phosphate coating methods.,* in *Bioceramics,* P. Ducheyne, Christiansen, D.,, Editor. 1993, Butterworth-Heinemann Ltd: Philadelphia. p. 165-171.

10. Guipont, V., et al., *High-pressure plasma spraying of hydroxyapatite powders.* Materials science and engineering, 2002. **A325**: p. 9-18.

11. Yamada, K., et al., *Bone bonding behaviour of the hydroxyapatite containing glass-titanium composite prepared by the Cullet method.* Biomaterials, 2000. **22**: p. 2207-2214.

12. Kumar, R.R. and M. Wang, *Functionally graded bioactive coatings of hydroxyapatite/titanium oxide composite system.* Biomaterials, 2001.

13. Kay, J.F., *Hydroxylapatite-coatings for non-precision implant treatments,* in *X-ray diffraction method for the quantitative characterisation of calcium phosphate coatings,* E. Horowitz and E. Parr, Editors. 1994. p. 149-162.

14. Tong, W., et al., *Studies on diffusion maximum in x-ray diffraction patterns of plasma-sprayed hydroxyapatite coatings*. Journal of Biomed Mater Res, 1998. **40**: p. 407-413.

15. Khor, K.A., H. Li, and P. Cheang, *Significance of melt-fraction in HVOF cpray hydroxyapatite particles, splats and coatings*. Biomaterials, 2004. **25**: p. 1177-1186.

16. Gu, Y.W., Loh, N.H., Khor, K.A., Tor, S.B., Cheang, P., *In Vitro studies of plasma-sprayed hydroxyapatite / Ti-6AL-4V composite coatings in simulated body fluid (SBF)*. Biomaterials, 2003. **24**: p. 1603-1611.

17. Liu, D.M., Troczynski., Tseng., W.,, *Water-based sol-gel synthesis of hydroxyapatite: process development*. Biomaterials, 2001. **22**(13): p. 1721-1730.

18. Manso, M., et al., *Electrodeposition of hydroxyapatite coatings in basic conditions*. biomaterials, 2000. **21**: p. 1755-1761.

19. Mavropoulos, E., et al., *Dissolution of calcium deficient hydroxyapatite synthesised at different temperatures*. Materials Characterisation, Article in Press.

20. Liu, D.M., T. Troczynski, and W.J. Tseng, *Water-based sol-gel synthesis of hydroxyapatite : process development*. Biomaterials, 2001. **22**(13): p. 1721-1730.

21. Osborn, J.F. and H. Newesely, *The material science of calcium phosphate ceramics*. Journal of Biomaterials, 1980. **1**: p. 108-111.

22. Jou, Z.C., H.M. Chou, and D.M. Lui, *Hydroxyapatite coating via amorphous calcium phosphates*, in *In Biomedical Materials Research in the Far East*, X. Zhang and Y. Ikada, Editors. 1993, Kobunshi KanKokai: Kyoto, Japan. p. 117-118.

23. Barrere, F., et al. *In vitro dissolution of various calcium phosphate coatings on Ti6AL4V*. in *13th International symposium on ceramics in Medicine*. 2000. Bologna, Italy: Trans Tech Publications.

24. Chow, L.C., *Dissolution studies of Calcium Phosphates*. Journal of the ceramics society, Japan, 1991(99): p. 954-964.

25. Leadley, S.R., et al., *Investigation of the dissolution of the bioceramic hydroxyapatite in the presence of titanium ions using ToF-SIMS and XPS*. Biomaterials, 1997. **18**(4): p. 311-316.

26. Kokubo, T., et al., *Solutions able to reproduce in vivo surface changes in bioactive glass-ceramic A-W*. Journal of Biomedical Materials Research, 1990. **24**: p. 721-734.

27. Rogers, K.D., S.E. Etok, and R. Scott, *Structural characterisation of apatite coatings*. Journal of Materials Science, 2004. **ARTICLE IN PRESS**

28. Williamson, G.K. and H. W.H, *Microstrain and crystallite size determination*. Acta Metalurgica, 1953. **1**: p. 22.

29. Roome, C.M. and C.D. Adam, *Rietveld analysis of calcium phosphate coatings*. Biomaterials, 1995. **16**: p. 691-696.

30. Nelson, D.G.A. and J.D.B. Featherstone, *Preparation, analysis, and characterisation of Carbonated Apatites*. Calcified Tissue International, 1982. **34**: p. S69-S81.

31. Le Geros, R.Z., B. Bonel, and R. Legeros, *Types of H20 in human enamel and in precipitated apatites*. Calcified Tissue Research, 1978. **26**: p. 111-118.

32. Weng, J., et al., *The role of amorphous phase in nucleating bone-like apatite on plasma sprayed hydroxyapatite coatingsin simulated body fluid*. Journal of materials science letters, 1997. **16**: p. 335-7.

33. Liu, X., Tao, S., Ding, C.,, *Bioactivity of plasma sprayed dicalcium silicate coatings.* Biomaterials, 2002. **23**: p. 963-968.

34. Weng, J., et al., *Formation and characteristics of the apatite layer on plasma-spray hydroxyapatite coatings in simulated body fluid.* Biomaterials, 1997. **18**: p. 1027-1035.

35. Liu, X., S. Tao, and C. Ding, *Bioactivity of plasma sprayed dicalcium silicate coatings.* Biomaterials, 2002. **23**: p. 963-968.

36. Gu, Y.W., K.A. Khor, and P. Cheang, *In vitro studies of plasma-sprayed hydroxyapatite/Ti-6Al-4V composite coatings in simulated body fluid (SBF).* Biomaterials, 2003. **24**(9): p. 1603-1611.

37. Weng, J., et al., *The role of amorphous phase in nucleating bone-like apatite on plasma sprayed coatings in simulated body fluid.* Journal of Materials Science Letters, 1997. **16**: p. 335-337.

38. Gross, K.A., et al., *In vitro changes of hydroxyapatite coatings.* Journal of Oral Maxillofacial Implants, 1997. **12**(5): p. 589-597.

39. Walton, A.G., *The formation and properties of precipitates.* 1967, New York: Interscience.

40. Queiroz, A.C., et al., *Dissolution studies of hydroxyapatite and glass-reinforced hydroxyapatite ceramics.* Journal of Materials Characterization, 2003. **ARTICLE IN PRESS**.

41. Panda, R.N., et al., *FTIR, XRD, SEM and solid state NMR investigations of carbonate-containing hydroxyapatite nano-particles synthesized by hydroxide-gel technique.* Journal of Physics and Chemistry of Solids., 2003. **64**: p. 193-199.

42. Park, E., R. Condrate, and D. Lee, *Infrared spectral investigation of plasma spray coated hydroxyapatite.* Materials Letters, 1998. **36**: p. 38-43.

43. Blakeslee, K.C. and R. Condrate. Journal of American Ceramic society, 1971. **54**: p. 559.

44. Levitt, S.R. and R. Condrate. American Mineral., 1970. **49**: p. 1562.

45. Klein, C.P.A.T. and K. de Groot, *A study of solubility and surface features of different calcium phosphate coatings in vitro and in vivo: a pilot study.*, in *Ceramics in substitive and reconstructive surgery*, P. Vincenzini, Editor. 1991, Elsevier: Amsterdam. p. 363-374.

46. Radin, S.R. and P. Ducheyne, *Plasma spraying induced changes of calcium phosphate ceramic characterisitics and the effect on in the in vitro stability.* Journal of materials science: Materials in medicine, 1992. **3**: p. 33-42.

47. Paschalis, E.P., et al., *Degradation potential of plasma-sprayed hydroxyapatite-coated titanium implants.* Journal of Biomedical Materials Research, 1995. **29**: p. 1499-1505.

HYDROXYLAPATITE – NANOPHASE ALPHA ALUMINA COMPOSITE COATINGS ON Ti-6Al-4V

Zafer Evis and Robert H. Doremus
Materials Science and Engineering, Rensselaer Polytechnic Institute, Troy, NY, 12180

ABSTRACT

Composites of hydroxylapatite (HA) and nanophase (n)-alpha-(α)-Al_2O_3 with 5wt% CaF_2 were coated on Ti-6Al-4V rods by cold pressing and sintered in air or argon atmosphere at 1100°C to improve the bonding between the metal and the ceramic. 25 and 40wt% α-n-Al_2O_3–5wt% CaF_2–HA composites gave the best results of strong bonding. 25wt% α-n-Al_2O_3–5wt% CaF_2–HA and Ti-6Al-4V showed the same thermal expansion coefficients. The thermal expansion coefficient of the 40wt α-n-Al_2O_3–5wt% CaF_2–HA composite was slightly smaller than that of Ti-6Al-4V, which created a compressive stress on the coating, resulting in improved bonding between the metal and the ceramic.

INTRODUCTION

Hydroxylapatite ($Ca_{10}(PO_4)_6(OH)_2$) has been widely used as a bulk implant in non-load bearing areas of the body and as coatings on implant metals. HA is a bioactive ceramic, which can bond to bone, because it is very similar to the mineral part of bone [1,2]. It has a fracture toughness of approximately 1 $MPa\sqrt{m}$ [3].

HA- alumina composites can be made more resistant to high temperature sintering by substituting small amounts of impurities in the HA phase such as Na^+, Mg^{2+}, CO_3^{2-}, CaF_2 or increasing the Ca/P ratio in HA [4,5].

Coatings of HA have been used in orthopedic and dental implants. For the best bonding, the HA phase should be more than 95 % of the coating [6]. HA coatings generally have a thickness of 50 to 200 μm [7]. The porosity of the HA coating should be minimized because highly porous coatings result in weak bonding [6].

HA is generally coated on metals by plasma spraying commercially [8]. In this technique, the spraying temperature of HA powders on metals exceeds HA's melting temperature (1670 °C). (α or β) tri-calcium phosphate (TCP), which is more degradable than HA, can form in addition to HA during plasma spraying. Some of the commonly used metals as coating substrates are commercially pure titanium, Ti-6Al-4V, 316L stainless steel, and CoCrMo alloys. Metals and HA have different thermal expansion coefficients (CTE) [9]. As a result of this difference, tensile or compressive forces result in cracks on HA coatings and poor adhesion of HA on the metal substrate. To overcome the difference between the CTE of HA and Ti-6Al-4V, HA-alumina composites were coated on Ti-6Al-4V.

EXPERIMENTAL PROCEDURES

The materials used in this research are pure HA, composites of HA and nanophase-α-alumina, and Ti-6Al-4V as substrate metal in coatings.

HA was synthesized by a precipitation method by mixing reagent grades of calcium nitrate and di- ammonium hydrogen phosphate in the alkaline pH region [10]. The powders of nanophase Al_2O_3, (48nm particle size, Nanophase Technologies Inc., Burr Ridge, IL) were mixed with HA powder. Nanophase γ-Al_2O_3 (as received) was transformed to nanophase α-Al_2O_3 by a heat treatment in a platinum crucible at 1300 °C for 10 minutes before mixing with HA.

The compositions of the coatings used in this research were 10, 25, and 40wt% α-n-Al$_2$O$_3$ – HA with 5wt% CaF$_2$. The dried HA particles were ground to $\leq$75 μm (-200 mesh) powder using a mortar and pestle, and calcined at 900°C for 1 hour. The calcined HA and nanophase α-Al$_2$O$_3$ powders were mixed by ball milling. The ball milled and dried powders were ground with a mortar and pestle.

The composites were coated on Ti-6Al-4V rods with a diameter of 4.75 mm. The composite powder was cold pressed at ~100 MPa around the metal to produce a total diameter of 12.7 mm. The coatings were sintered in air or argon at 1100°C for 1hr.

The bond strength of the coatings on metal rods was determined by a push-out test. The ceramic coating and the metal substrate were pushed away from each other using an INSTRON Universal Testing Machine. A speed of 0.1 mm/min was applied during the push-out test. The following formula was used to calculate the push-out strengths of the coatings.

$$S = \frac{P}{(\pi * d * L)} \qquad (1)$$

where P: applied load, d: metal rod diameter (4.75 mm), L: height of the ceramic coating along metal rod

The CTE of the composite coatings and Ti-6Al-4V was measured with an Orton automatic dilatometer from RT to 1100°C. Samples were cut to cylindrical shapes of 1" length and 0.2" diameter. Samples were heated in the dilatometer at a rate of 3°C/min.

RESULTS and DISCUSSION

X-ray diffraction results of HA– α-n-Al$_2$O$_3$ composites with 5wt% CaF$_2$, which were sintered at 1100°C, are presented in Figure 1. The HA and α-n-Al$_2$O$_3$ were stable in the presence of CaF$_2$ after sintering at 1100°C. CTE decreased as the amount of Al$_2$O$_3$ increased. The CTE of the metal and the ceramic coatings are presented in Figure 2. CTE of Ti-6Al-4V is close to that of the composite containing 25wt% α-n-Al$_2$O$_3$ and 5wt% CaF$_2$.

Sintering of the coatings in air resulted in excessive oxidation of the metals. This oxide formation on the metal surfaces diminished the bonding between the metal and the coating. Bond strengths of HA and 5wt%CaF$_2$- α-n-Al$_2$O$_3$ - HA coatings on Ti-6Al-4V by push out tests are presented in Figure 3. Increasing the amount of nanophase-Al$_2$O$_3$ in the coatings improved the bonding strength because the CTE of Ti-6Al-4V and HA-alumina coatings were more closely matched. The composite containing 40wt% α-n-Al$_2$O$_3$ and 5wt% CaF$_2$ resulted in highest bond strength between the metal and ceramic because of the compressive stresses formed on ceramic. To make HA and nanophase-Al$_2$O$_3$ composites, α-n-Al$_2$O$_3$ powder was used because of its better sinterability than γ-n-Al$_2$O$_3$ [11].

Diffusion and mass transport affect the final densification by controlling the removal of pores and increasing the reactivity of the matrix and particles. For example, Ca^{2+} ions were transported from CaF$_2$ or HA to the contact surface of Al$_2$O$_3$ particles. At temperatures above about 1150°C in ambient atmosphere, HA begins to decompose by the following reaction:

$$Ca_{10}(PO_4)_6(OH)_2 \rightarrow 3Ca_3(PO_4)_2 + CaO + H_2O \qquad (2)$$

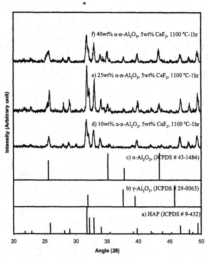

Figure 1: X-ray diffraction spectra of 10, 25, and 40wt% α-n-Al₂O₃ – HA composites with 5wt% CaF₂ sintered at 1100°C

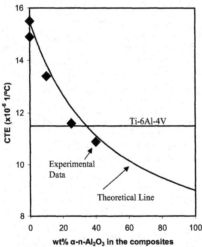

Figure 2: Coefficients of thermal expansion of HA matrix composites with α-n-Al₂O₃

CaF₂ prevented reaction 2 so that water is not given off and fewer pores are formed. The addition of calcium fluoride to the powder mixtures of the composites strongly reduced the tendency of the HA to decompose during sintering. Presumably the fluoride substitutes for OH⁻ groups in the apatite structure; the reaction can be written as:

$$Ca_{10}(PO_4)_6(OH)_2 + CaF_2 \rightarrow Ca_{10}(PO_4)_6F_2 + CaO + H_2O \qquad (3)$$

The reaction between HA and calcium fluoride can occur at temperatures (900°C) lower than the sintering temperature of 1100°C [12]. Therefore this reaction eventually improved the stability of the composites at 1100°C.

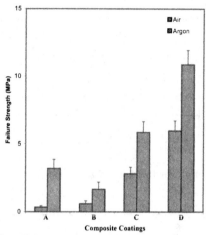

Figure 3: Failure strengths of the bond between the Ti-6Al-4V substrates and the ceramic coatings of A: HA; B: 10wt% α-n-Al₂O₃, 5wt% CaF₂, -HA; C: 25wt% α-n-Al₂O₃, 5wt% CaF₂, -HA, D: 40wt% α-n-Al₂O₃, 5wt% CaF₂, -HA after the push-out test

Coating of HA on Ti-6Al-4V is difficult because of their different thermal expansion coefficients, which results in crack formation. When the CTE of the coating and the metal is matched, no crack formation can be observed, as seen in Figure 4. Otherwise, cracks generally start to form at the interface between the metal and ceramic and result in cracking of the ceramic coating. Measured CTE's of the composites agreed closely with CTE's from a simple mixture formula, as shown in Figure 3. The formula is [13]:

$$\alpha_C = \frac{\alpha_m E_m V_m + \alpha_p E_p V_p}{E_m V_m + E_p V_p} \qquad (4)$$

where E and V represent the Young's modulus and the volume fraction of the matrix (m), the particles (p), and α the CTE.

Pure HA or 10wt% α-n-Al₂O₃-HA with 5wt% CaF₂ composite resulted in poor bonding with the titanium alloy because their CTE was higher than that of the alloy, resulting in a tensile stress on the coating. The CTE of 25 and 40wt% α-n-Al₂O₃- HA with 5wt% CaF₂ composites matched with that of Ti-6Al-4V, giving better bonding of the coating on the metal by eliminating tensile stresses between them.

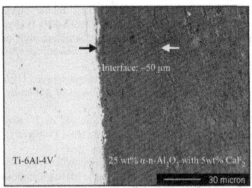

Figure 4: Optical microscopy image 25 wt% α-n-Al$_2$O$_3$-5wt% CaF$_2$ composite coated on Ti-6Al-4V, no crack formation by matching the CTE of the coating and the metal

CONCLUSION

Composites of HA and nanophase-α-Al$_2$O$_3$ with 5wt% CaF$_2$ were coated on Ti-6Al-4V rods by cold pressing and sintering. Matching the thermal expansion coefficient of the ceramic coating and the metal substrate improves the bonding and minimizes the formation of cracks. Nanophase alumina–HA coatings showed better bonding than the pure HA coatings on Ti-6Al-4V. Addition of nanophase alumina into HA resulted in smaller thermal expansion coefficient, bringing it close to that of Ti-6Al-4V.

ACKNOWLEDGEMENTS

The authors would like to thank Dr. Rich Sarno of CeramTec North America Corp., New Lebanon, NY, for the coefficient of thermal expansion measurements of the materials used in this study.

REFERENCES

[1] R. B. Martins, M. W. Chapman, N. A. Sharkey, S. L. Zissimos, B. Bay, E. C. Shors, "Bone Ingrowth and Mechanical Properties of Coralline Hydroxyapatite 1 Year After Implantation," *Biomaterials*, 14 341-45 (1993).

[2] Z. Jianguo, Z. Xingdong, C. Muller-Mai, U. Gross, "The Early Host and and Material Response of Hydroxyapatite / Beta-tricalciumphosphate Porous Ceramics After Implantation into the Femur of Rats," *Journal of Materials Science: Materials in Medicine*, 5 243-51 (1994).

[3] M. B. Thomas and R. H. Doremus, "Fracture Strength of Dense Hydroxylapatite," *American Ceramic Society Bulletin*, 60 [2] 258-9 (1981).

[4] A. C. Tas, "Synthesis of Biomimetic Ca-Hydroxyapatite Powders at 37 °C in Synthetic Body Fluids," *Biomaterials*, 21 1429-38 (2000).

[5] H. Kim, Y. Koh, S. Seo, H. Kim, "Properties of Fluoridated Hydroxyapatite-Alumina Biological Composites Densified with Addition of CaF$_2$," *Materials Science and Engineering C*, 1072 1-7 (2003).

[6] K. Soballe and S. Overgaard, "The Current Status of Hydroxyapatite Coating of Prosthesis," *The Journal of Bone and Joint Surgery*, 78-B 689-91 (1996).

[7]   S. R. Sousa and M. A. Barbosa, "Effect of Hydroxyapatite Thickness on Metal Ion Release from Ti6Al4V Substrates," *Biomaterials*, **17** 397-404 (1996).

[8]   R. Y. Whitehead, W. R. Lacefield, L. C. Lucas, "Structure and Integrity of a Plasma Sprayed Hydroxylapatite Coating on Titanium," *Journal of Biomedical Materials Research*, **27** 1501-7 (1993).

[9]   Y. C. Yang and E. Chang, "Residual Stress in Plasma-sprayed Hydroxyapatite Coating Measured by the Material Removal Method," *Journal of Materials Letters*, **22** 919-22 (2003).

[10]   M. Jarcho, C. H. Bolen, M. B. Thomas, J. Babock, J. F. Kay, R. H. Doremus, "Hydroxyapatite Synthesis and Characterization in Dense Polycrystalline Form," *Journal of Materials Science,* **11** 2027-35 (1976).

[11]   S. Chang, "Sintering of Nanophase Alumina Powder and Alumina Composites Reinforced with C-Nanotubes and Creep of Ceramic Matrix Composites," Ph. D. Thesis, Rensselear Polytechnic Institute, Troy, NY, (2000).

[12]   J. C. Elliott, "Structure and Chemistry of the Apatites and Other Calcium Orthophosphates"; Elsevier, NY, 161-286 (1994).

[13]   R. Warren and V. K. Sarin, "Particulate Ceramic-matrix Composites"; in *Ceramic Matrix Composites* Edited by R. Warren, Chapman and Hall, NY, 1992.

A SYSTEM FOR *IN VIVO* ANCHORING OF IMPLANTS TO HARD TISSUE

L. Hermansson,
Uppsala University
Department of Materials Science
Box 534
751 21 Uppsala
Sweden

A. E. Åbom
Linköping University
Thin Film Physics Division
681 83 Linköping
Sweden

H. Engqvist, J. Loof and N. Axén
Doxa AB
Axel Johanssons gata 4-6
SE-754 51 Uppsala
Sweden

ABSTRACT

To allow for early loading of an implant in bone tissue and to reduce the risk for long-term loosening, high quality early fixation is crucial. Even small gaps may lead to micromotions between the implant and the tissue, which increases the risk of fibrous tissue formation and eventually implant loosening. Porosity or cavities in the tissue surface also reduce the implant fixation. The materials system investigated in this paper aims at providing a system for *in vivo* anchoring of an implant to bone tissue, by using coatings of chemically bonded ceramic materials (CBC-materials). Implants coated with CBC ceramics may also favorably be combined with cement pastes of the same type of material, to react together *in vivo*.

Chemically bonded ceramics therefore potentially provides systems for fast *in vivo* anchoring of implants to bone tissue characterized by: a high strength bond between the coating to the implant surface, resulting from the build-up of a chemical bond as the coating cures; inter-connection of sub-layers of the coatings by their co-hydration; and for selected applications anchoring of the CBC coating to a CBC paste by co-hydration. The dissolution-precipitation mechanisms of CBC's also contribute to the anchoring of the ceramic to the biological tissue.

This paper presents results from a first investigation on an implant fixation system based on coatings of a high strength chemically bonded ceramic, calcium aluminate. Evidence is presented of coating densification and anchoring mechanisms of the coating to a metal surface, realized by precipitation of hydrates.

INTRODUCTION

The use of implants made from artificial materials has become a very established method of restoring function to damaged parts of the human body, particularly for hard tissues. Materials for implants have to fulfil a variety of characteristics including, biocompatibility, strength, flexibility, chemical resistance, etc. Only a selection of materials has been proven capable of matching the requirement profile. For bone contact established biomaterials are: Stainless steel, pure or alloyed titanium, Co-Cr-alloys, some ceramic materials such as zirconia and alumina, and polymers, mainly polymethylmetacrylates [1].

In orthopaedics, it is difficult to achieve both initial strength of the bone-implant interface, while also maintaining long-term function. Often the creation of interfacial stability fails due to lack of biocompatibility or insufficient early stabilisation of the implant. For successful implantation early stabilisation is crucial. Even small gaps may lead to relative movements, micro-motions, between implant and the tissue, which increases the risk of implant loosening over time due to formation of zones of fibrous tissues at the implant-tissue interface. Production of wear debris and mismatch of elastic modulus are other reasons to failed implantation [1].

One relatively established method to reduce the problem with poor interfacial stability for implants, is surface coatings technology. With coatings technology, structural characteristics of the implant (e.g. strength, ductility, low weight or machinability) may be combined with surface properties promoting tissue integration [2-3]. There are several established coating deposition techniques, e.g. Chemical Vapour Deposition, Physical Vapour Deposition, Thermal Spray Deposition and Electrolytic Deposition [4].

Bioactivity is the characteristic of a material to promote the formation of a bond between an artificial material and living tissue [5]. Promotion of strong bond between the implant and tissue reduces many drawbacks related to the interfacial stability. Only a very small group of materials can claim bioactivity. Calcium phosphates, including hydroxyapatite are today the most prominent member of the group of bioactive materials, together with bio-glasses (phosphorous containing glasses) [6]. Also various surface modifications via oxidation and roughness may improve the implant bonding to bone tissue [1].

Calcium phosphates are also deposited as coatings, most commonly with spraying techniques. Today plasma sprayed hydroxyapatite represents the most commonly used coating type for orthopaedic implants. It is a fairly young type of product; the first reported clinical results of hydroxyapatite coatings on hip prosthesis are from the 1980's [1]. Since then, both temporary and permanent hydroxyapatite coated implants are used for implants such as screws, spinal components and various joint prostheses.

However, this type of coatings has also produced failures resulting from: mechanical weakness of the coating material or the coating/implant interface, lack of uniformity of morphology and crystallinity, increased dissolution leading to production of debris. These problems sometimes cause inflammation and implant loosening [1].

This paper presents initial results on the development of implant fixation system for hard tissue based on coatings of chemically bonded ceramics (CBC). There are several chemically bonding ceramic systems, mainly calcium based; e.g. Ca-silicates, Ca-aluminates, Ca-sulphates and Ca-phosphates. The chemistry of these materials is similar to that of hard tissue in living organisms, the latter being based on apatites and carbonates. Chemically bonded ceramics can be bioactive and generally display great potential as biomaterials [7-10]. This paper focuses on calcium aluminate, and investigates possible coating deposition techniques [11, 12] for this material as means of achieving better system for implant fixation.

MATERIALS AND METHODS
Plasma spraying
Coatings were produced with plasma spraying following a similar procedure as for hydroxyapatite coatings. Substrates of stainless steel (316) were sand blasted with alumina grit of 250 μm grain size to a surface roughness of about 2 μm. Crystalline calcium aluminate of the fast hydrating $(CaO)_{12}(Al_2O_3)_7$ phase was produced by sintering and crushing. The powder was sieved to irregular grains of 50-90 μm in size. The powder was plasma sprayed to coating thickness of about 70 μm using Ar as carrier gas. Most coatings were deposited in a 350A/67V plasma with the gun 70 mm from the substrate surface, but both current, voltage and substrate distance was

varied for evaluation purposes. During deposition the substrates reached about 70 °C, and was left to cool before being removed. Coatings were also produced on titanium grade 5 substrates with similar results. The substrates were cleaned ultrasonically in acetone and alcohol before deposition.

A selection of plasma sprayed coatings was stored for 48 hrs in 100% relative humidity of water at 37 °C to study the effects of hydration. The plasma sprayed coatings were evaluated with scanning electron microscopy (SEM) and X-ray diffraction (XRD) before and after hydration. Fracture surfaces parallel to the substrate were achieved for microscopy purposes in a tear test.

Sputtering

Sputtered films were deposited in a UHV system with a background pressure of 10^{-8} Torr using a radio-frequency power source and powder feeds around 100 W. A planar $(CaO)_{12}(Al_2O_3)_7$ calcium aluminate target was used. The target was sintered from crushed and sieved powder of about 10 μm grain size. The films were grown on planar substrates of Si (polycrystalline) and Ti (grade 2). The target was directly facing the substrates, and a target-to-substrate distance of 7 cm was used. The sputtering pressure was 8 mTorr in most cases.

The coatings were deposited at temperatures ranging from ambient temperature to 460 °C. The deposition temperatures were determined by mounting a thermocouple on the surface of the substrates. No in-situ preparation of the substrates, such as pre-sputtering, was performed. After the deposition, the samples were cooled down to ambient temperature, before being removed from the vacuum system.

Prior to deposition all substrates were cleaned by a standard process of 10 minutes in Decon in an ultrasonic bath, followed by 10 minutes in acetone and 10 minutes in isopropanol and then immediately transferred into the load-lock of the deposition system. The samples were blown dried in dry N_2 gas and mounted on the sample holder. The sputtered coatings were evaluated with scanning electron microscopy (SEM) with Energy Dispersive X-ray spectrometry (EDX), transmission electron microscopy (TEM) and X-ray diffraction (XRD).

RESULTS

Plasma spraying

Plasma spraying produced coatings of calcium aluminate with an appearance very similar to plasma sprayed calcium phosphate coatings, see Fig 1. The splat structure is clearly seen on top-sections, including typical features such as internal pores, cavities and cracks. The features are confirmed in cross-sections together with a generally good apposition to the substrate. Initial coating adhesion tests lead to cohesive fractures at about 65 MPa.

X-ray analysis revealed a reduced crystallinity for the coatings as compared to the powder (100% crystalline). The crystallinity was roughly 30-40 % for the 350A/67V, 70 mm distance parameters, but varied clearly with the power of the plasma – a high energy reduced the cristallinity. The crystalline phases are the same in coating as in the precursor powder.

After hydration for 48 hrs in water-saturated air, the microstructure had clearly altered. A large share of the calcium aluminate is hydrated and in larger cavities hydrate crystallites can be identified with microscopy. This is illustrated in Fig 2 for the cleavage surfaces resulting from tear testing orthogonal to the coatings surface.

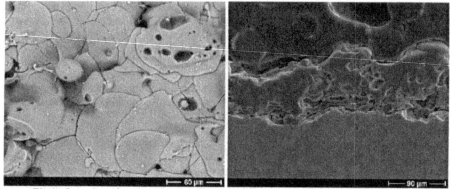

Fig.1. Scanning electron microscopy images of a non-hydrated plasma sprayed calcium aluminate coating as a top view (left) and in cross-section (right).

Back-scatter scanning electron images (SEM) of cleaved non-hydrated coatings reveals the fresh cleavage planes as lighter areas (Fig. 2 right), and porosities as darker (deeper down) areas. The same microscopy on hydrated coatings (Fig. 2 left) reveals cleavage planes with hydrated (darker) areas and still non-hydrated material (lighter areas), together with extensively grown hydrate crystallites filling previous pores and cavities.

Fig. 2. Scanning electron microscopy images of cohesive fractures parallel to the coating resulting from tear testing directed orthogonal to the surface of (left) a hydrated and (right) a non-hydrated plasma plasma sprayed calcium aluminate coating.

Sputtering

Radio-frequency sputtering with a calcium aluminate target produced even and dense calcium aluminate films with a satisfactory adhesion to the substrate, both for Ti and Si substrates. The produced films consist of Ca, Al and O as measured by EDX, but with a deficiency of Ca and an enrichment in O as compared to the target composition, for all deposition temperatures, see Table I. The film composition corresponds well to the $(CaO)(Al_2O_3)$ calcium aluminate phase.

The calcium aluminate films appear dense as observed with TEM. There is no apparent grain structure, and the border between calcium aluminate and substrate is hardly distinguishable, see Fig. 3.

Table I. Composition determined by EDX of target and deposited films.

	Target $C_{12}A_7$	Deposited at 170 °C	Deposited at 330 °C	Deposited at 460 °C
Ca	23 %	14 %	13 %	14 %
Al	27 %	25 %	28 %	27 %
O	50 %	61 %	59 %	59 %

The films were amorphous as measured by X-ray diffraction (X-ray amorphous) for the lower deposition temperatures used. For coatings deposited at 460 °C some X-ray reflections are seen. The growth rate is about 60 nm/hour. With electron diffraction in a TEM no diffraction pattern was shown for any of the films, possibly due to a very small grain size.

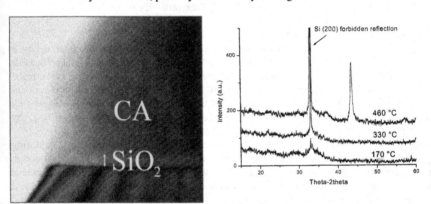

Fig. 3. Transmission electron microscopy image of the transition zone between calcium aluminate coating and a silicon substrate for a film grown at 330 °C (left) (the arrow indicates the SiO_2 layer and is 10 nm long), and X-ray diffraction patterns (right) of samples grown on Si substrates for temperatures between 170 and 460 °C.

DISCUSSION
Plasma spraying
        This work shows that calcium aluminate can be plasma sprayed under normal spraying conditions. The achieved films have a similar appearance as apatite coatings deposited with the same technique. It is also found that the sprayed coatings hydrate in water rich environments, but at a much lower rate than for a calcium aluminate powder of the same phase. This is believed to be a result of the high density and lower surface area of the sprayed coating compared to a powder.

        After 48 hrs at body temperature in a saturated water environment, a significant hydration had taken place, which densified the coating. There was however, still non-hydrated material in the structure, and still significant porosity.

Potentially, a plasma sprayed ceramic coating can be significantly densified through hydration after deposition. In an *in vivo* situation, the formed hydrates would precipitate both against the substrate and against a surrounding hard tissue wall.

Sputtering

Radio frequency sputtering produced X-ray amorphous coatings below about 460 °C. This is accordance with sputtering of alumina, which does not generally result in crystalline films at temperatures below 500 C°. The growth rate of calcium aluminates is relatively low. Four hours deposition, resulted in about 250 nm thick films. A decreased target to substrate distance increases the deposition rate, as does an increasing target power.

The plasma was found to be relatively unstable during deposition. At irregular times, the plasma shut off. Varying the power between 60 and 110 W did not solve the problem. Sputtering at 8 mTorr was more stable than both lower and higher pressures.

In all films a contamination of Ni was found. This contamination most likely originates from a clamping ring holding the target to place or from the earth shield.

Hydrating coatings as the basis for an implant fixation system

A system for fast *in vivo* anchoring of implants to bone tissue should be characterized by: a high strength bonding of the coating to the surface of the implant, for CBC materials achievable through pretreatments of the implant and formation of chemical bonds as the coating cures; inter-anchoring of individual sub-layers of the coatings achievable by liquid transport and co-hydration; for selected applications also anchoring of the CBC coating to a CBC based paste by co-hydration; as well as anchoring of the CBC ceramic to the biological tissue, e.g. by dissolution-precipitation mechanisms and volume increase.

This paper investigates the prospects of achieving these characteristics by exploring coatings of chemically bonded ceramics (CBC) on metal substrates. CBC materials offer through their curing characteristics based on hydration with water, unexplored possibilities to achieve high mechanical integrity of the coating and substrate-coating interface, and also to provide fast and high strength bonding to the hard tissue wall. CBC ceramics solidify through chemical reactions with water. For CBC ceramics, the high temperatures and pressures of traditional ceramics are not necessary; hence these materials can be made to solidify in-vivo while reacting with the water and ions of body fluids. This paper focuses on calcium aluminate as the coating material.

Calcium aluminate is generally used as a ceramic powder, which, when mixing with water, hardens through a chemical reaction, hydration. During the hydration a solid body of hydrates is formed. Calcium aluminate is characterised by an unusually high turn-over of water during hydration. The hardening process is chemically driven and demands no firing or high pressures. Due to the high amounts of water involved in the hydration, the material forms bodies of low residual porosity and improved mechanical strength.

An important aspect of chemically bonded ceramics including calcium alumiante is the interaction with body fluids during in-vivo hydration, which forces the hydrophosphates (HPO_4^{2-} and $H_2PO_4^-$) to transform into PO_4^{3-} ions. These ions together with Ca-ions and hydroxy-ions from the basic Ca-aluminate system immediately precipitate as apatite, $Ca_5(PO_4)_3(OH)$, due to the low solubility of the apatite phase [12]. As a consequence, calcium aluminate curing in a body fluid environment shows bioactivity.

The high turn-over of water during the hydration of calcium aluminate provides the basis for better mechanical integrity of a coating of this material, and for the formation of a rapidly developed bond to adjacent bone tissue. The hydration after deposition produces increased bonding to the substrate, better mechanical integrity of the coating, and built up a better bond to the bone tissue. When a coated implant is positioned in a bone cavity a mass increase will take place due to reactions with water in the body fluid. Instead of initial point contacts, the mass increase leads to enlarged contact zones, including formation of apatite towards the tissue. Thereby a coating system of the same principles as today's hydroxyapatite coatings is produced, but characterised by a higher degree in-vivo materials formation leading to higher reliability, early loading characteristics and reduced risk of loosening over time.

Also the chemical bond between the CBC coating and the implant surface can be enhanced by a chemical pretreatment of the surface from the original metallic or ceramic character to a double oxide of titanium interacting with the chemically hydrating ceramic.

Further, a CBC paste positioned between the coated implant and the bone tissue, and which co-hydrates with the CBC coating will have the function of filling larger gaps between the implant and the biological tissue, including vacuoles or cavities in the surface of the bone.

CONCLUSIONS

It is concluded that conventional plasma spraying can be used to deposit thick calcium aluminate coatings. The technique produces coatings of comparable appearance as hydroxyapatite coatings. The plasma sprayed calcium aluminate can be hydrated, and thereby forms hydrate crystallites throughout the coating thickness.

It is further concluded that conventional radio-frequency sputtering from a sintered calcium aluminate target can be used to deposit thin coatings of calcium aluminate. These films are X-ray amorphous at least below 460 °C. A good apposition to the substrate and coating density was achieved as revealed from TEM images.

To allow for early loading of an implant to bone tissue, the described coating system using *in situ* and *in vivo* hydration, based on Ca-aluminate provides an opportunity for improved fixation of implants including both chemical and biological integration.

ACKNOWLEDGMENTS

CAM Implants bv is acknowledged for deposition of the plasma sprayed coatings.

REFERENCES

[1]J.E. Ellingsen, S.P. Lyngstadaas (Eds.), "Bio-implant interface, improving biomaterials and tissue reactions", (2003) CRC Press LLC.

[2]S.Vercaigne et al, "Bone healing capacity of titanium plasma-sprayed and hydroxylapatite coated oral implants", *Clin. Oral Implants Res*, **9**, 261 (1998).

[3]X. Liu, C. Ding, Z. Wang, "Apatite formed on the surface of plasma sprayed Wollastonite coating immersed in simulated body fluid", *Biomaterials*, **22** (2001) 2007-2012.

[4]D.S. Rickerby, A. Matthews (Eds.), "Advanced Surface Coatings", (1991) USA:Chapman & Hall, New York.

[5]D.F. Williams, Definitions in Biomaterials, Progress in Biomedical Engineering, Vol 4, (1987)

[6]L.L. Hench, "Biomaterials: a forecast for the future", *Biomaterials*, **19** (1998) 1419-1423.

[6]H. Engqvist, J-E. Schultz-Walz, J. Loof, G. A. Botton, D. Maye , M. W. Pfaneuf, N-O.Ahnfelt, L. Hermansson, "Chemical and biological integration of a mouldable bioactive ceramic material capable of forming apatite in vivo in teeth", *Biomaterials* vol **25** (2004) pp. 2781-2787.

[7]L. Hermansson, L. Kraft, H. Engqvist. Chemically bonded ceramics as biomaterials, *Key Engineering Materials* vol. **24** (2003) pp. 437-442.

[8]L. Kraft, Ph D Thesis, Calcium Aluminate Based Cement as Dental Restorative Materials, Dec 2002, Uppsala University

[9]N. Axén, T. Persson, K. Björklund, H. Engqvist, L. Hermansson, An injectable Bone Void Filler Cement Based on Ca-Aluminate, *Key Engineering Materials* Vols. **254-256** (2004), pp. 265-268.

[10]J.M. Schneider, S. Rohde, W. D. Sproul and A. Matthews, "Recent developments in plasma assisted physical vapour deposition", *J. Phys. D: Appl. Phys.* **33** (2000) R173-R186.

[11]L. Ericksson, R. Westergård, N. Axén, H. Howthorn, S. Hogmark, Cohesion in plasma-sprayed coatings - a comparison between evaluation methods, *Wear* **214** (1998) 30-37.

[12]J. Samachson, Nature 1968; 218:1261.

Dental Materials and Applications

APATITE FORMATION ON A BIOMINERAL BASED DENTAL FILLING MATERIAL

H. Engqvist, J. Loof, L. Kraft, L. Hermansson
Doxa AB
Axel Johanssons gata 4-6, Se-754 51 Uppsala, Sweden

ABSTRACT

The objective of the study was to investigate if apatite is present on the surface of a biomineral (calcium aluminate) based dental filling material *in vivo*. The ability to form apatite *in vivo* reduces the risk of interfacial instability with the host tissue. If the material forms apatite continuously a high-quality sealing of a cavity would be possible and also a reduction of the wear rate.

An extracted tooth with a two-year-old medium sized occlusal-distal biomineral based filling (calcium aluminate) was analyzed using XPS. Reference spectra from enamel were obtained. The peaks for Ca and P on enamel were compared with the corresponding peaks for the mineral filling.

The XPS-spectra showed that the elements C, O, N, Ca, P, Si and Al were present on the surface. The same elements were present on enamel except from Al. The location of both the Ca (2p3) and P (2p) peaks were the same for both the biomineral and enamel. The results prove that in vivo, upon the biomineral-filling surface, both Ca and P are present and bonded very similar to Ca and P in enamel. The energy shifts for Ca and P correspond to apatite according to reference literature.

INTRODUCTION

One of the major reasons for replacing a tooth restoration is interface instability, i.e. secondary caries [1]. With time the interfacial instability leads to gap formation between the restoration and the tooth where bacteria will enter and, as result, secondary caries will develop. This is especially a problem for composite fillings, which shrink upon hardening and therefore need to be bonded to the tooth. The bonding may fail due to the internal stresses built up, and a gap is produced. One route to solve the problem would be to develop a material that is bioactive, i.e. form a bond directly to the tooth [2]. This is a difficult task since the material also needs to fulfill many other requirements regarding strength and biocompatibility to function as a restorative material. One obvious way to achieve a true chemical bonding to tooth substance would be to use calcium phosphate to fill the cavity. But the amount of bonded water in calcium phosphate cements is very low. To achieve a high strength low porosity material very little water can be used, which results in a non-injectable consistency [3]. Therefore much effort has been put on the development of a composite material with light activated monomers and calcium phosphates [4]. Another route could be to use chemically bonded minerals. Some chemically bonded minerals are known to bond much water during hydration, e.g. the calcium silicates and the calcium aluminates [5]. Especially a material based on minerals from the calcium aluminate system has a good potential, due to high initial strength and rapid hardening [6,7]. The material can be delivered as a powder that is mixed with water, accelerator and minuscule amounts of a dispersing agent to an injectable paste. During hydration all the added water is consumed and to fill porosities some extra water is absorbed via the body fluid. Furthermore, the calcium aluminate has been proven to be bioactive *in vivo* by formation of a tight bond to the tooth by apatite formation at the interface and *in vitro* by formation of apatite on the surface in simulated body fluid [8,9]. In the *in vivo* study, fillings were made in wisdom teeth that were extracted after 4 weeks, cut and the tooth-filling interface was analysed with various analytical techniques (SEM, TEM, FIB, electron diffraction and EDS). From the study (and from other studies) a question can be raised if apatite is formed upon free surfaces in simulated body fluid should not also apatite be present on the tooth surface *in vivo*. If it were so, the chemistry would then lead to a continuous new formation of apatite on the surface, which would then indicate a "self-healing" property and continuous seal of the cavity.

MATERIALS AND EXPERIMENTAL
In this study a restorative material based on calcium aluminate minerals was used [6, 7]. The minerals were compressed to tablets, which upon mixing with water were condensed and then hardened via a reaction between the powder and the water. Due to the high amounts of water involved in the hydration, a high degree of mouldability is achieved, and a dense body of low residual porosity and high strength is formed [5]. At mixing with water the minerals dissolve and new minerals precipitate according to:

$$3\ CaO \bullet Al_2O_3\ +\ 12\ H_2O\ \rightarrow\qquad Ca_3[Al(OH)_4]_2(OH)_4\ +\ 4Al(OH)_3 \qquad (1)$$

| Marokite | water | Katoite | Gibbsite |

The precipitated minerals (hydrates) have a very fine grain size and a high surface energy, and the amount of bonded water is considerable. The hydrates bond the material together and strength develops. The compressive strength reaches values close to 200 MPa and the flexural strength about 60 MPa [6].

A tooth with a medium sized occlusal distal filling was extracted due to odontological reasons. It was carefully ground on the opposite unfilled mesial surface. It was ground plane parallell and could then be more easily handled. The distal surface was analysed with X-ray photoelectron spectroscopy (XPS). The instrument used was a PHI Quantum 2000 X-ray with a spot size of 100μm from monochromatized Al kα radiation. The XPS technique is a very surface sensitive technique (depth resolution of 100 Å) but with a low lateral resolution. The strength of the technique is that it can determine energy shifts for different atomic states. There are reference spectra for all elements in the periodic table and their respective energy shifts when bonded to other elements, e.g. the energy shift for the Ca2p3 peak between elemental Ca and when bonded to O in CaO.

On the overview spectra the pass energy was 93.9 eV and step size (= energy resolution) 0.8 eV. Both Ca and P peaks were analyzed more carefully with pass energy 58.7 eV, step size 0.5 eV. Reference spectra from enamel were also obtained. The peaks for Ca and P on enamel were compared with the corresponding peaks for the biomineral filling material.

RESULTS
The spectra from the biomineral showed that the elements C, O, N, Ca, P, Si and Al were present on the surface, see Fig. 1a. The same elements were present on enamel except from Al. The shift of both the Ca (2p3) and P (2p) peaks were the same for both the biomineral material and enamel, see Fig. 1b and c.

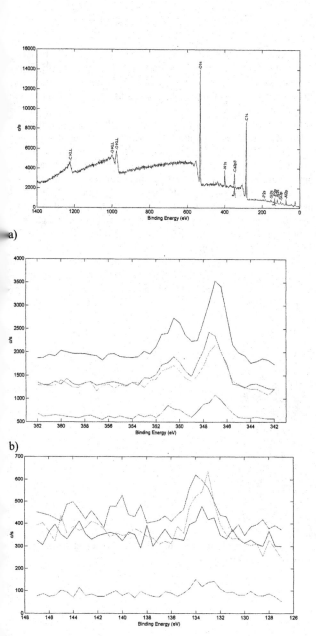

Fig. 1. Overview spectra from the biomineral filling a) and the comparative Ca and P spectra b) and c) respectively. The three uppermost curves in b) and c) were from the biomineral and the bottom curve was from enamel.

DISCUSSION

The results prove that on the biomineral surface both Ca and P are present and bonded similar to Ca and P in enamel. The energy shifts for Ca and P correspond to apatite [10]. This strongly indicates that an apatite layer is formed on top of the calcium aluminate material not only *in vitro* but also *in vivo*. A restorative material with the ability to constantly form an apatite film could seal the cavity preventing microleakage.

To further prove the marginal seal of the cavity the filling surface could also be analysed with XRD and SEM combined with EDS. Also reference spectra for calcium aluminate and katoite should be obtained and compared to the spectra in Fig. 1b. This could give valuable information about the difference in Ca bonding between the minerals. Unfortunately the high vacuum destroyed the tooth when going up to normal air pressure. More extracted teeth are thus needed for a complete battery of analyses. It is also necessary to study the occlusal surface to see if the apatite formation is continuous also on surfaces exposed to wear. As already been found in vitro, the presence of P on the surface and the identical shifts for Ca and P compared to enamel gives a strong indication that apatite precipitate on the calcium aluminate surface also in vivo.

The formation of apatite on biomineral materials can be understood from considering the respectively solubility products of the Ca-hydrates formed [8]. Logically apatite has a low solubility at high pH and forms readily. The other Ca-hydrates that can form (e.g. katoite) have a higher solubility and therefore do not form if also phosphate ions are present. A typical apatite layer formed on calcium aluminate is seen in Figs. 2 and 3.

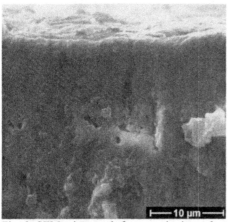

Fig. 2. SEM micrograph from apatite layer formed on a calcium aluminate specimen formed *in vitro* after storage in phosphate buffer solution for 4 months.

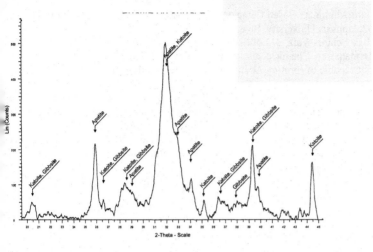

Fig. 3. X-ray diffraction pattern from apatite layer formed on a calcium aluminate specimen formed *in vitro* after storage in phosphate buffer solution for 4 weeks.

CONCLUSIONS

The biomineral restorative material bonds Ca and P on its surface *in vivo* with the same energy shifts as on enamel. This inherent property can have a major impact on the marginal integrity of a restoration leading to a better seal of the cavity and thus reduction of the incidence of micro-leakage.

REFERENCES

1. I. A. Mjör, J. E. Moorhead, Reasons for replacement of restoration in permanent teeth in general dental practice. International Dental Journal 2000;50:361-366.
2. L. L. Hench, Biomaterials: a forecast for the future. Biomaterials 1998;19:1419-1423.
3. M. Nilsson, Injectable calcium sulphate and calcium phosphate bone substitutes, Ph. D thesis, Lund (2003).
4. M. S. Park, E. D. Eanes, J. M. Antonucci, D. Skrtic. Mechanical properties of bioactive amorphous calcium phosphate/metacrylate composites. Dental Materials 1998;3;14:137-141.
5. Lea's Chemistry of Cement and Concrete. Edited by P.C. Hewlett, Arnold 1998.
6. J. Loof, H. Engqvist, K. Lindqvist, N-O. Ahnfelt, L. Hermansson, Mechanical properties of a permanent dental restorative material based on calcium aluminate, Journal of Materials Science: Materials in Medicine, Vol. 14, No. 12, (Dec 2003), 1033-1037.

7.   L. Kraft, Calcium Aluminate Based Cement as Dental Restorative Materials, Ph D Thesis, December 2002, Uppsala University, Sweden.

8.   H. Engqvist, J-E. Schulz-Walz, J. Lööf, G. A. Botton, D. Mayer, M. W. Phaneuf, N-O. Ahnfelt, L. Hermansson: Chemical and biological integration of a mouldable bioactive ceramic material capable of forming Apatite in vivo in teeth, Biomaterials, 25 (2004) 2781-2787.

9.   O. Seung-Han, C. Se-Young, L. Yong-Keun, N. K. Kyoung, C. Seong-Ho, Effects of litium fluoride and maleic acid on bioactivity of calcium aluminate cement: Formation hydroxyapatite in simulated body fluid, J Biomed Mater Res, 67A-104-111 (2003).

10.   Takadama et. al., An X-ray photoelectron spectroscopy study of the process of apatite formation on bioactive titanium metal, Journal of Biomedical Materials Research, 55,(2) 2001; May2001, pp. 185-193.

CHEMICALLY BONDED CERAMICS WITH AN ADDITIONAL ORGANIC BINDING SYSTEM

H. Engqvist, H. Spengler and N-O Ahnfelt
Doxa AB,
Axel Johanssons gata 4-6,
Uppsala, Sweden

L. Hermansson
Materials Science Department,
The Angstrom Laboratory,
Uppsala University, Sweden

ABSTRACT

This paper describes a system for bioactive chemically bonded ceramic (CBC) materials, having improved properties related to rheology as well as end product properties related to the mechanical profile. Two binding systems - an inorganic system based on Ca-aluminates (CA), and an organic system based on poly acrylic acids - are activated in the over-all setting/hardening process.

The two binding phases may work at separate time and also over-lapped in the process facilitating the combination of early improved mouldability with high performance end features mainly related to compression and bend strength and a visco-elastic behaviour. The basic CA system neutralises the initial acidity in the polyacrylic system. The viscosity of the material can be controlled within wide frames upon initial mixing of the powdered material and the hydration liquid, from moist granules to an injectable slurry.

INTRODUCTION

For materials, such as dental filling materials and implants, that are to interact with the human body, it is an advantage that the materials are made as bioactive or biocompatible as possible. Other properties that are required for dental filling materials and injectable implants are good handling ability with simple applicability in a cavity, moulding that permits good shaping ability, hardening/solidification that is sufficiently rapid for filling work without detrimental heat generation and provides serviceability directly following therapy, chemical stability, good bonding between the biomaterial and the biological wall, dimensional stability, radio-opacity, good long time properties and good aesthetics especially regarding dental filling materials.

To meet such advanced property profiles it is almost impossible to rely on just one material class or system, e.g. a chemically bonded ceramic, a glass ionomer cement, a resin based polymer or other separate system. This paper describes a novel chemically bonded system comprising at least two bonding systems, one Ca-aluminate based system and one organic, acid based system. Description of the Ca-aluminate system is presented elsewhere [1-5]. To use this system aiming at high strength materials requires powder compacts with high green density, which is a problem in moulding the material. The two binding systems presented in this paper may work at separate time and also over-lapped in the process facilitating the combination of potential early-age properties with high performance including appropriate consistency related to mouldability and end features especially related to biomechanical and biochemical properties. The reaction mechanisms and the over-all behaviour will be discussed and screening tests and preliminary data will be presented.

MATERIALS

Raw materials

The CBC material used in this work is based on a mono calcium aluminate, Marokite, $(CaAl_2O_4)$, with a maximum particle size of approximately 10 μm, produced at Doxa. As filler two different dental glasses and fumed silica were used, Table 1.

Table 1. Constituents of screening powder blends. Physical data from manufacturers.

Material	Particle size μm	Density g/cm³	Index of refraction n_d	Abbreviation
Ca-aluminate	10.5	2.98		CA
Dental glass 1	1.5	3.42	1.60	G1
Dental glass 2	0.4	3.00	1.55	G2
Ordinary Portland cement	10-12	3.18		OPC
Fumed silica	0.014	2.31	1.46	μ-SiO₂

The organic compounds used are compiled in Table 2. All chemicals used were bought from Sigma-Aldrich.

Table 2. Compilation of chemicals used in the experiments.

Material	Molecular weight (Mw)	pH of 10 % w/w solution in water at 23°C	Abbreviation
L(+) tartaric acid	150.1	1.44	TTA
Citric acid	210.1	1.82	CitA
Poly acrylic acid (l)	2,000	2.18	PAA
Poly(acrylic-co-maleic acid) (l)	3,000	1.63	PAMA1
Poly(acrylic-co-maleic acid) sodium salt (s)	50,000	7.87	PAMA2
Poly(acrylic-co-maleic acid) sodium salt (s) lightly crosslinked [1]	50,000	-	PAMA3
Isopropanole p.a. quality	60.1	-	IPA

[1] Super-absorbent

Powder preparation

The sample constituents were blended in a 250 mL polyethylene bottle, with 125 g silicone nitride balls as milling media (diameter 8 mm) and isopropanole, IPA. The bottle was put on a roller for 3 hours. The slurry was evaporated in a Laborota 4003-digital roll evaporator at 130 mbar, 50°C and 50 rpm for 2h. The moist powder was dried at 80°C in an oven for another 2 hours. Note that the drying temperature should not exceed the glass temperature, T_g, of the polymer (i.e. 106°C for polyacrylic acid [6]). The dry powder was stored in polyethylene bottles at room temperature.

The ceramic powders were either mixed with the polymer by hand or by adding the polymer directly in the powder blend. If the polymer was added as a liquid it was added to the powder with a 50 µL Hamilton syringe prior to the mixing in a mixing machine (ESPE RotoMix, 3M). The mixture was blended between 6 and 10 seconds in the RotoMix, depending on how easily mixed it was.

Polymer milling and preparation

The PAMA 2 and PAMA 3 particles were approximately 1 mm in diameter as received from the manufacturer. The particle size was reduced by additional milling using silicone nitride balls (ϕ = 8 mm) in IPA. The polymer was weighed to approximately 30 g and put in a polyethylene bottle together with the balls and IPA. The bottle was put on a roller for 24 h. The polymer was dried stored in a glass beaker in an oven at 80°C until no IPA was left. The particle size was measured in a Malvern Mastersizer from Malvern Instruments. The D_{99} for the particles was approximately 22 µm.

Storage media

The test specimens were stored at 37°C in a humid environment. The specimens were placed on a grid above distilled water in a plastic container with lid (the relative humidity was close to 100%).

EVALUATION TECHNIQUE AND TEST METHODS

Evaluation technique

In this work a feature called contour plot in MODDE 6.0 [7] was used for guidance to see trends in the interacting factors of interest (consistency, hardness and flexural strength). The contour plot has the structure as followed in Fig. 1.

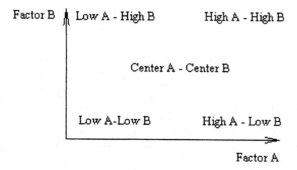

Figure 1. Map of interacting factors A and B in a contour plot.

In the contour plot the responses (the measured results) are displayed in the plot giving the reader an overview in which direction a certain property is enhanced. The process factors variables in the different experimental designs were amount of polymer added, TTA, OPC and w/c (w/p)-ratio (water to cement ratio and water to powder), see Table 3 for the whole test program. Selected data are presented due to limited space in this presentation. All work sheet data were fit with Partial Least Square method, PLS [7].

Table 3. Boundaries for statistical experimental design.

MODDE trial	w/c	w/p	PAMA 2 (%-Weight)	TTA (%-Weight)	OPC (%-Volume)
1	0.31 – 0.85	0.20 – 0.40	5 – 30	0 –10	-
2	0.31 – 0.56	0.20 – 0.30	5 - 20	0 - 10	-
3	0.39 – 0.45	0.24 – 0.28	10 - 17	10	0 - 4
4	0.39 – 0.49	0.24 – 0.30	5 - 10	10	-

The model validity is represented by R^2 and Q^2 varying from 0 to 1 and - ∞ to 1, respectively. R^2 represents the 'goodness of fit' and is an estimate of how good the regression model can be fit to the raw data. R^2 equal to 1 represents a perfect model and 0 no model at all. Q^2 is used for indication of the model validity and represents the 'goodness of prediction'. A Q^2-value above 0.5 corresponds to a good model and above 0.9 excellent. For a model to pass this diagnostic validity test, the R^2 and Q^2 values should not differ more than 0.2 – 0.3 [7].

Test methods

Consistency

All compositions were graded on a scale from 1 to 5 with regard to their consistency appearance (i.e. 1=dry, powder, 2=moist granules, 3=moist ball or paste, 4=almost flowing, 5=flowing), see Fig. 2.

Class 1   Class 2   Class 3   Class 4   Class 5

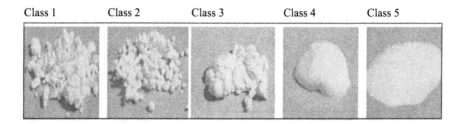

Figure 2. Classification of consistency. Examples from left to right - dry to flowing.

Micro hardness

The micro hardness was tested with a Buehler micro hardness tester using a load of 100 g. The samples were condensed in an acrylic cavity (diameter 3.5 mm) with dental instruments and stored at 37°C in a humid environment. Before measuring the micro hardness the samples were polished down to a fineness of 4,000 grit with a silicone nitride grinding paper. Micro hardness was measured as Vicker hardness three times on each specimen (a total of 4 specimens) and calculated as a mean value (i.e. from a total of 12 measurements).

Flexural strength

The biaxial flexural strength test was based on the ASTM F-394 standard. The specimens were produced by condensing the material into polyurethane plates (5 mm in diameter and 4 mm deep). The samples were pressed out into a container and stored at 37°C when hydrated. The raw specimens were then stored at 37°C in a humid environment. The specimens were polished down to 0.7 mm with a 623 Disc Grinder from Gatan Inc. for controlled parallel grinding and polishing.

The circular specimen was placed on three balls positioned on the periphery of an imagined circle of a specified diameter. The samples in the ASTM F-294 standard are 3.175 cm (1.25 inches) but due to specimen preparation conditions, these samples are 5 mm in diameter. The minimum thickness of the specimen must be such that its centre of deflection at failure does not exceed one-half of the specimen thickness. An experimentally decided thickness of 0.7 mm was therefore chosen.

The specimens were loaded with a fourth ball with a diameter of 1.6 mm and pressed upon until failure in a Zwick test rig. The maximum load of five specimens per sample was registered and re-calculated to a mean value in MPa using the ASTM F-394 equations.

RESULTS
Pre-screening test - consistency

Initially hands-on tests to screen the consistency possibilities were conducted on different powder compositions, see Table 4. The w/w, w/p and and w/c ratios describe different aspects of liquid and added polymer content to the total mount of powder used.

Table 4. Compilation of boundaries in the screening of workable compositions.

Test	Powder	PAMA 1 w/w[1]	PAMA 2 w/w[1]	PAMA 3 w/w[1]	w/p[2]	w/c[3]	Liquid
1	CA	0.05- 0.16	-	-	0.16 – 0.40	0.16 - 0.40	Water
2	CA	-	0.00 – 0.12	-	0.20, 0.30, 0.40	0.20, 0.30, 0.40	10 % TTA
3	CA	-	-	0.00 – 0.12	0.20, 0.30, 0.40	0.20, 0.30, 0.40	10 % TTA
4	CA and G1 (70/30 v/v)	-	0.05, 0.175, 0.30	-	0.20, 0.30, 0.40	0.31, 0.43, 0.63, 0.85	0 – 10 % TTA, 7.5 Li
5	CA and G1 (70/30 v/v)	-	0.05, 125, 0.20	-	0.20, 0.25, 0.30	0.31, 0.37, 0.47, 0.56	0 – 10 % wt TTA

[1] w/w is weight of added polymer divided by total amount powder.
[2] w/p is calculated as the liquid added divided by total amount of powder
[3] w/c is calculated as the liquid added divided by total amount of cement (CA or CA and OPC).

The pure CA was used in the tests 1-3, see Table 4 for lower and upper composition limits and various polymers used. In the test the w/c-ratio was varied from 0.16 to 0.40 and the amount of the different types of polyacid, i.e. PAMA 1, PAMA 2 and 3 varied from 0.01 to 0.16 w/w. In these tests water and a 10 % wt solution of TTA were used. The choice of using TTA comes from the glass ionomer cement (GIC) theory described elsewhere [8]. In the pre-screening no effect or influence from the TTA could be seen but was not discarded due to the fact that many tests had been done with the addition of TTA.

Regarding the composition of the CA powder and PAMA 1 and 2 one could clearly see that the consistency gets more fluent with an increase of water and/or polymer, examples given in Tables 5-6. Above a w/c-ratio of 0.30 the CA and PAMA 1 material was fluent. It was also noted that the blend was slow setting.

Table 5. The influence of w/c-ratio and PAMA 1 on consistency with CA powder in water.

w/w polymer	w/c					
	0.16	0.20	0.25	0.30	0.35	0.40
0.05	-	1-	2-	4	5	5
0.10	-	2+	4	4+	5	5
0.16	3	3	4	5	5	5

Table 6. The influence of w/c-ratio and PAMA 2 on consistency with CA powder in water.

w/w polymer	w/c		
	0.20	0.30	0.40
0.00	1	2+	4
0.01	2	3-	4
0.02	2	3-	4
0.04	2	3-	5
0.08	2	3-	5
0.12	2-	3-	5

In the case with CA and PAMA 3 (lightly cross-linked), it was observed that the cross-linked PAMA 3 absorbed water, giving an average of '2' in consistency. The consistency was spongy throughout the test series. The PAMA 3 is a so-called super-absorbent, and was excluded from further testing.

Screening tests
For guidance in interpreting the results of consistency, hardness, and flexural strength a contour plot was used in MODDE 6.0 [7]. A screening test with CA and G1 in proportion 70/30 (vol./vol.) and PAMA 2 for 24 h was conducted. The reason for introducing the glass phase is to obtain more practical study cases. Based on that, and on the decision to use 35 % vol./vol. of G1 on behalf of CA for increased radio-opacity, an even more narrow trial was performed with 5, 7.5 and 10 % wt PAMA 2 in a CA and G1 premixed powder blend (65/35 vol./vol.), see Figs. 3-4.

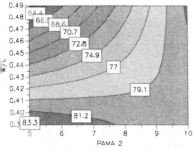
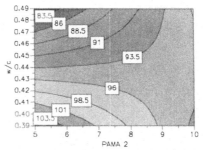

Figure 3. Screening of hardness (7 d),
CA and G1 powder with PAMA 2 (wt-%)
(Q^2=0.238, R^2=0.722)

Figure 4. Screening of hardness (28 d)
CA and G1 powder blend with PAMA 2 (wt-%)
(Q^2=0.297, R^2=0.816)

The compiled results of the hardness as a function of time are shown in Fig. 5. The hardness showed a maximum for the low polymer sample with 5 % wt PAMA 2 at a w/c-ratio of 0.39. The behaviour of increased hardness with time was seen in all samples.

Micro hardness,HV

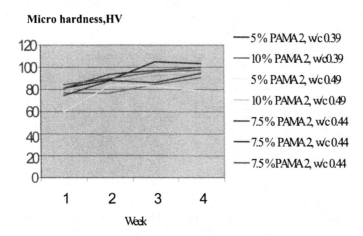

Figure 5. Vickers hardness measured at 1, 2, 3 and 4 weeks.

The compositions were also tested for flexural strength. The results of the flexural strength measurements at 7 days and 28 days, Figs. 6-7, showed the need for lower polymer content and lower w/c-ratios.

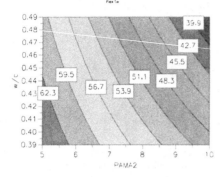
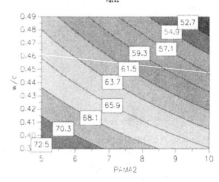

Figure 6. Screening of flexural strength, 7d
CA and G1 powder and PAMA 2 (wt-%)
(Q^2=0.617, R^2=0.989)

Figure 7. Screening of flexural strength, 28d
CA and G1 powder and PAMA 2 (wt-%)
(Q^2=0.319, R^2=0.810)

Further testing

An inert glass named G2 with a particle size of 0.4 μm in a 65/35 (vol./vol.), was tried in addition to glass G1. The influence of w/c-ratio and PAMA 2 on consistency with CA and G2 powder blend in water, and the influence of w/c-ratio and PAMA 2 on the hardness, HV (24 h) was evaluated. The results are summarised in Table 7.

Table 7. The influence of w/c-ratio and PAMA 2 on consistency and hardness for CA and G2 powder.

w/w polymer	Consistency scale		HV 24 h	
	w/c0.31	w/c0.37	w/c 0.31	w/c 0.37
0.020	-	4+	-	74.8
0.024	2+	4+	81.6	66.7
0.026	2+	4+	76.5	96.2

DISCUSSIONS AND OUTLOOK

An experimental version of the two-system material is being evaluated with reference to general mechanical properties, and typical data are presented in Table 8. This material is based on mono-Ca-aluminate, a dental glass G2 and a Na-salt of the polyacrylic acid. The particle size of the main constituents is given in Fig. 8

Table 8. Selected properties of a two-phase system. Standard deviation within brackets.

Property	Data	
Consistency	5	Scale in Fig. 2
Setting time	5	min
Compression strength, 24 h	145 (15)	MPa
Compression strength, 1 month	225 (40)	MPa
Flexural strength, 24 h	45 (7)	MPa
Flexural strength, 1 month	60 (11)	MPa
Hardness, 24 h	80	Hv
Hardness, 1month	120	Hv
Young's modulus	25	GPa
Linear expansion	0.15	%

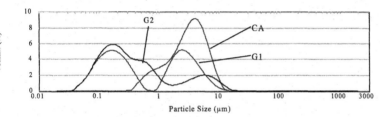

Figure 8. Particle size distribution of pure CA, G1 and G2, respectively.

A follow-up study will focus primarily on the initial strength aspects, especially the property development during the first hours. This is important in achieving early loading capacity for the two-system material as an *in vivo* injected material [9].

CONCLUSION

It is feasible to implement poly(acrylic-co-maleic acid) and its salts into the CBC system to yield a two-system material characterized by a flowing cement system with appropriate end point mechanical properties. In the MODDE trials the R^2 and Q^2-values were close to diagnostic validity but not perfect, indicating a complex relationship between the process factors. An increased polymer content and amount of liquid reduce the mechanical properties. Tartaric acid does not seem to influence consistency, and the effects of this acid on the mechanical strength are yet unknown.

REFERENCES

[1] Hermansson, L, Kraft, L, Engqvist, H. Chemically bonded Ceramics as Biomaterials, Proceedings of the 2nd International Symposium on Advanced Ceramics, Key Engineering Materials, vol 24 7, 2003.

[2]   Kraft L. Calcium Aluminate Based Cement as Dental Restorative Materials, Ph.D. Thesis, Uppsala, 2002.
[3]   K. Scrivener, A. Capmas, Calcium Aluminate Cements, Lea's Chemistry of Cement and Concrete. Edited by P.C. Hewlett, Arnold 1998.
[4]   J. Loof, H. Engqvist, K. Lindqvist, N-O. Ahnfelt, L. Hermansson, Mechanical properties of a permanent dental restorative material based on calcium aluminate, Journal of Materials Science: Materials in Medicine, Vol. 14, No. 12, (Dec 2003), 1033-1037.
[5]   H. Engqvist, J-E. Schulz-Walz, J. Lööf, G. A. Botton, D. Mayer, M. W. Phaneuf, N-O. Ahnfelt, L. Hermansson: Chemical and biological integration of a mouldable bioactive ceramic material capable of forming Apatite in vivo in teeth, Biomaterials, 25 (2004) 2781-2787.
[6]   Polymer Handbook, 4th Ed.
[7]   L. Eriksson, E. Johansson, N. Kettaneh-Wold, C. Wikström, S. Wold. Design of Experiments - Principles and Applications. Umetrics Academy IID 1016, 2000
[8]   Nicholson JW, Brookman PJ, Lacy OM et al. Fourier transform infrared spectroscopic study of the role of tartaric acid in glass ionomer dental cements. J Dent Res 1988;67:1451-4.
[9]   Doxa AB, Technical Report, Early Loading of Implants, March 2004 (in preparation)

FRACTURE PROPERTIES OF CORTICAL BONE AND DENTIN

J. J. Kruzic, R. K. Nalla and R. O. Ritchie*
Materials Sciences Division
Lawrence Berkeley National Laboratory, and
Department of Materials Science and Engineering
University of California, Berkeley, CA 94720

J. H. Kinney
Lawrence Livermore National Laboratory
Livermore, CA 94550

ABSTRACT

In order to assess the fracture properties of cortical bone and dentin, *in vitro* R-curve experiments were performed on human cortical bone (34 – 99 year-old humeri) and elephant dentin specimens hydrated in Hanks' Balanced Salt Solution. Experiments yielded toughness values very similar to ceramics, with R-curves rising over the range of 1 – 3.5 MPa√m for dentin and 2 – 5 MPa√m for human cortical bone; such R-curve behavior was attributed in both cases to the formation of bridging ligaments in the crack wake. Crack bridges were observed both by microscopy and X-ray tomographic techniques, and the extent and nature of the bridging zones were quantitatively characterized using compliance based experiments. In contrast to ceramics, both of these materials exhibit viscoplastic (creep) deformation behavior at room temperature, which leads to considerable time-dependent crack tip blunting. Additionally, time-dependent cracking behavior (static fatigue) was observed in both cortical bone and dentin; however, for dentin, crack arrest occurred quite quickly due to the rapid rate of crack blunting. Finally, the effects of dehydration and age on the fracture properties were also considered with the goal of gaining a micro-mechanistic understanding of the resistance to fracture in these structural biomaterials.

INTRODUCTION

Cortical bone and dentin are two of the most abundant naturally occurring mineralizing tissues in the body, and accordingly, a thorough understanding of their fracture properties on a mechanistic level is needed in order to assess *in vivo* fracture risk, develop treatments to reduce the effects of aging and disease, and to successfully design implant materials such as bioceramics. Both of these tissues are composed of similar building blocks, namely collagen fibers, which consist of aligned collagen molecules, impregnated with nanometer scale carbonated apatite crystals. At larger size scales, however, the microstructures of these tissues are quite different. While dentin is composed of a scaffold of collagen fibers, with cylindrical tubules (~ 1 – 2 μm wide) running from the dentin-enamel junction to the inner pulp of the tooth,[1] cortical bone has a hierarchical microstructure, consisting of collagen fibers arranged into bundles and aligned in lamellae (~ 3 – 7 μm thick), which layer in concentric rings to form osteons roughly 10 – 500 μm in diameter; the osteons have large vascular channels, the Haversian canals (up to 50 – 90 μm diameter), running down the center.[2] Dentin and cortical bone, therefore, essentially provide two different microstructures of the same nominal material (i.e., mineralized collagen); the present mechanistic characterization of their fracture properties is intended to provide insights into how microstructural features affect their resistance to fracture.

* Corresponding author. Tel: +1-510-486-5798; fax: +1-510-486-4881. *E-mail address:* roritchie@lbl.gov (R. O. Ritchie)

EXPERIMENTAL PROCEDURES

Materials

Dentin from fractured shards of elephant tusk from an adult male elephant (*Loxodonta africana*), and fresh frozen human cadaveric humeral cortical bone from nine donors (34 – 99 years old - cause of donor death unrelated to skeletal state) were used in this study. Blocks of bone were obtained by carefully sectioning the medial cortices of the mid-diaphyses of the humeri. To measure the toughness, compact-tension, C(T), specimens were used. Specimens had thicknesses of $B \sim 1.2 - 3.3$ mm, widths of $W \sim 13 - 18.3$ mm and initial notch lengths of $a_o \sim 3.1 - 5.5$ mm. Dentin samples were orientated such that the nominal crack growth direction was perpendicular to the long axis of the tubules and the crack plane was in the plane of the tubules. Cortical bone samples were divided into three age groups- arbitrarily named *Young*, *Middle-Aged* and *Elderly*. These samples were orientated with the starter notch and the nominal crack-growth direction along the proximal-distal direction, i.e., parallel to the long axis of the osteons and hence, long axis of the humerus, with the crack in the longitudinal-radial plane (i.e., C-L orientation[3]). All specimens were ground and polished, finishing with a 0.05 µm alumina suspension. Razor-micronotches (root radius of ~15 µm) were placed at the end of saw-cut notches by repeatedly sliding a razor blade over the saw-cut notch while irrigating with a 1 µm diamond slurry. Specimens were kept hydrated throughout preparation and prior to testing; additional details may be found eleswhere.[4,5]

R-curve testing

R-curves were measured to evaluate the resistance to fracture in terms of the stress intensity, K, as a function of crack extension, Δa, under a monotonically increasing driving force. After thoroughly hydrating specimens in Hanks' Balanced Salt Solution (HBSS) for at least 40 hours at room temperature in air-tight containers, tests were conducted in ambient air (25°C, 20 – 40% relative humidity) with the specimens being continuously irrigated with HBSS. The specimens were loaded in displacement control using standard servo-hydraulic testing machines (MTS 810, MTS Systems Corporation, Eden Prairie, MN) with a (quasi-static) loading rate of ~0.015 mm/s (quasi-static) until the onset of cracking, at which point the sample was unloaded by 10 – 20% of the peak load to record the sample load-line compliance at the new crack length using either a linear variable-displacement transducer mounted in the load frame or an external displacement gauge mounted on the specimen grips. This process was repeated to determine fracture resistance, K_R, as a function of crack extension, Δa. Crack lengths, a, were calculated from the compliance data obtained during the test using standard C(T) load-line compliance calibrations.[6] Due to crack bridging, errors invariably occurred in the compliance crack lengths; accordingly, re-calibration to the actual crack length was periodically achieved using optical microscopy. Differences between the compliance and optically measured crack lengths were corrected by assuming that any such error accumulated linearly with crack extension. In some cases, dentin specimens were tested in a dehydrated state in a rough vacuum chamber (~0.1 Pa). Full details on testing procedures may be found in Refs. 4,5.

Crack Path Determination

After R-curve testing, crack paths were examined using optical microscopy and scanning electron microscopy (SEM) in order to observe the interaction of cracks with microstructural features and the relevant toughening mechanisms. To determine the behavior below the sample surface, synchrotron X-ray computed tomography was performed on two bone and two dentin specimens at the Stanford Synchrotron Radiation Laboratory (SSRL), Menlo Park, CA. Imaging was performed with monochromatic 25 keV X-rays, with a voxel size (spatial resolution) of ~5 μm and exposure times of ~10 minutes for every 1 mm of crack length. Further details of this technique are described elsewhere.[7]

Bridging Zone Characterization

To quantitatively assess the extent and nature of the bridging zone behind the crack tip in cortical bone, a multi-cutting compliance technique similar to that of Wittmann and Hu[8] was used for three post R-curve specimens (37 – 41 year old). Specifically, a diamond saw blade was employed to sequentially cut out the crack, and thus incrementally eliminate the crack wake (in steps of ~0.25 – 1 mm), while the sample compliance was measured after each incremental saw cut. When the portion of the crack wake that is eliminated is traction-free, there is no change in compliance after cutting, i.e., the saw cut notch behaves as a bridge-free crack. However, if active bridges are eliminated from the crack wake by the saw blade, a corresponding increase in the sample compliance is expected. By using this technique, the bridging zone length, L, can be assessed by noting the notch length when the sample compliance begins to increase. Furthermore, a normalized bridging stress distribution can be estimated from the multi-cutting compliance data using:[8]

$$\frac{\sigma_{br}(x)}{\sigma_{max}} = \frac{C^2(a)C_m'(x)}{C_m^2(x)C'(a)} ,$$ (1)

where a is the crack length, C is the ideal, bridge-free, compliance, $C_m(x)$ is the measured compliance after cutting to the position x, measured from the load line, and $C'(z) = dC(z)/dz$. The bridging stress distribution, $\sigma_{br}(x)$, obtained from the multi-cutting compliance data is normalized by the factor σ_{max}, which corresponds to the maximum bridging stress at the crack tip. The ideal compliance may be computed using standard compliance calibrations.[6]

Alternatively, one may assume a commonly used bridging function a priori in order to estimate the bridging stress distribution,[9-13] viz:

$$\frac{\sigma_{br}}{\sigma_{max}} = (1 - \frac{X}{L})^n ,$$ (2)

where X is the distance behind the crack tip (X = $a - x$), and n is an exponent that describes the shape of the bridging stress distribution as it decreases from σ_{max} at the crack tip to zero at X = L. The exponent, n, may be determined from the multi-cutting compliance data using the relation:[11]

$$\frac{L}{n+1} = \frac{C(a)}{C'(a)}(\frac{C(a)}{C_m(a)} - 1) ,$$ (3)

where all symbols have been previously defined.

In order to further quantify the bridging stresses, it must first be noted that the bridging stress distribution, σ_{br}, may be related to the bridging stress intensity, K_{br}, by the relationship:[14,15]

$$K_{br} = \int_0^a h(a,x)\sigma_{br}dx ,$$ (4)

where x is the position measured from the load line, a is the crack length, and $h(a, x)$ is a geometry-dependent weight function, derived for the C(T) geometry to be:[15]

$$h = \sqrt{\frac{2}{\pi a}} \frac{1}{\sqrt{1-x/a}} [1 + \sum_{(v,\mu)} \frac{A_{v\mu}(a/W)^{\mu}}{(1-a/W)^{3/2}} (1-x/a)^{v+1}] .$$ (5)

The coefficients $A_{v\mu}$ may be found in ref. 15. By combining Eqs. 1, 4 and 5, or Eqs. 2, 4, and 5, the unknown σ_{max} may be found for each proposed bridging function provided the bridging stress intensity, K_{br}, is known.

Time-dependent cracking experiments

In many materials, cracks can propagate subcritically, i.e., at stress intensities less than the critical value, K_c, under the action of sustained loads. In order to assess whether dentin or cortical bone experiences time-dependent cracking, C(T) specimens were tested under constant displacement conditions at 37°C while immersed in HBSS. The specimens were loaded under displacement control until the R-curve initiation toughness was slightly exceeded, after which the load-line displacement was held fixed. During the test, the load was monitored as a function of time, with a drop in load indicating that either crack extension or creep deformation (causing crack blunting) had occurred. In order to distinguish between the two, the test was periodically interrupted in order to make direct observations of the crack using optical microscopy. For the case of crack extension with minimal creep, the load versus time data may be converted into crack length versus time data using compliance calibrations, assuming a linear compliance curve with fixed origin. Corrections were made for discrepancies between the actual and compliance determined crack lengths, as described above.

RESULTS

R-curve results

R-curves for cortical bone and for dentin are shown in Fig. 1a and 1b, respectively. Cracks were grown between 5 and 7 mm during the course of each test. For bone, the R-curves rose linearly with mean slopes of 0.39 (S.D. = 0.09), 0.16 (S.D. = 0.06), and 0.07 (S.D. = 0.03) MPa√m/mm for the *Young*, *Middle-Aged*, and *Elderly* groups, respectively. The *crack-initiation*

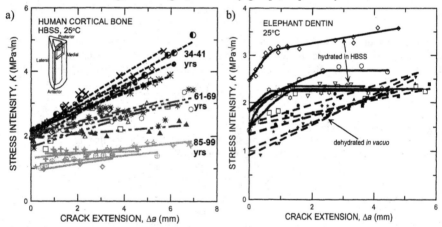

Figure 1. R-curves for (a) hydrated human cortical bone and (b) hydrated (open symbols) and dehydrated (solid symbols) elephant dentin.

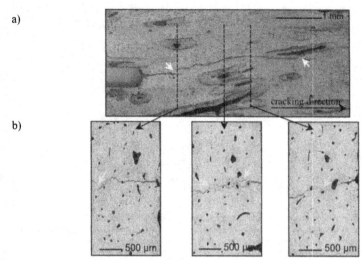

a)

b)

Figure 2. a) Optical micrograph and b) X-ray tomographic reconstructions of crack bridging (white arrows) in human cortical bone. In b), through-thickness slices are seen for locations marked in a).

toughness, K_o, was obtained by extrapolating a linear fit of the data for each sample to $\Delta a = 0$, with mean K_o values for the *Young, Middle-Aged*, and *Elderly* groups of 2.03 (S.D. = 0.19), 1.96 (S.D. = 0.18), and 1.22 (S.D. = 0.20) MPa√m, respectively. For dentin, the R-curves for the dehydrated state rose linearly (slope 0.26 ± 0.05 MPa√m/mm), similar to bone, while hydrated dentin displayed an initially steeper slope (0.54 ± 0.16 MPa√m/mm) followed by a plateau region (of slope 0.06 ± 0.04 MPa√m/mm), where the toughness remained essentially constant with crack extension. Furthermore, the dehydrated specimens, showed a lower crack-initiation toughness, $K_o = 1.18 ± 0.20$ MPa√m, as compared to 1.88 ± 0.40 MPa√m for the hydrated specimens.

Crack bridging

Fig. 2a shows an example of typical crack growth from the notch in a hydrated cortical bone specimen. The discontinuous nature of the crack path indicates formation of uncracked ligament bridges (indicated by white arrows) behind the crack tip as well as "out-of-plane" deflection of the crack. On the sample surface, such ligaments, some as large as hundreds of micrometers in size, can be readily seen to bridge the crack. Furthermore, the X-ray tomography results in Fig. 2b confirm that such uncracked ligaments occur throughout the bulk of the specimens, and are not an artifact of the surface stress-state. Fig. 2b shows three reconstructed through-thickness (two-dimensional) slices at different crack lengths in the same specimen imaged optically in Fig. 2a; the existence of uncracked ligaments (as indicated by white arrows) throughout the bulk of the material is apparent. Additionally, it is seen that the crack typically does not penetrate the osteons. Indeed, the path taken by the crack for this orientation appears to be dictated by the *interface* of the osteonal system with the surrounding matrix, *i.e.* the cement line.

In hydrated dentin, although clearly delineated bridges were observed on the sample surface (Fig 3a), none were seen on the sample surface for dehydrated dentin (Fig. 3b). In both cases, however, X-ray tomography revealed definitive sub-surface bridging (Figs. 3c, 3d).

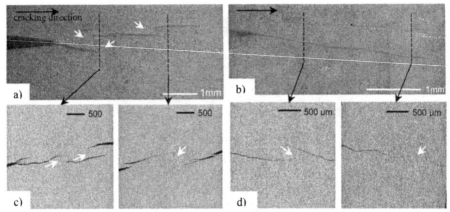

Figure 3. Optical micrographs of crack paths in a) hydrated and b) dehydrated dentin. X-ray tomographic reconstructions of through through-thickness slices are seen in c) and d) for hydrated and dehydrated dentin, respectively, revealing sub-surface bridging (white arrows) in both cases. In c) and d), through-thickness slices are seen for locations marked in a) and b).

Bridging Zone Characterization

Two methods, as described above, were used to estimate bridging stresses in three post R-curve cortical bone samples (Fig. 4). For these estimates, K_{br} was deduced from the R-curves, assuming:

$$K_{br} = K_{app} - K_o, \qquad (6)$$

where K_{app} is the applied stress intensity needed for crack extension, and K_o is the initiation toughness. Based on these values for K_{br}, σ_{max} for the three samples was found to range between 7 – 11 MPa for the stress functions given by Eq. 1, and 10 – 17 MPa for the

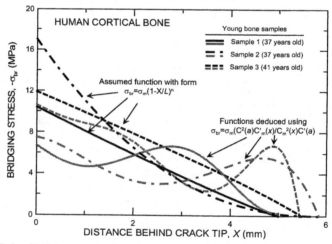

Figure 4. Deduced bridging zone stress distributions for three post R-curve human cortical bone samples.

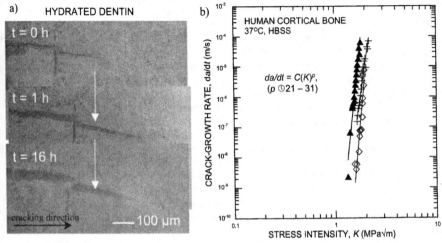

a) HYDRATED DENTIN

t = 0 h

t = 1 h

t = 16 h

cracking direction

——100 μm

b)

HUMAN CORTICAL BONE
37°C, HBSS

$da/dt = C(K)^p$,
$(p \oslash 21 - 31)$

STRESS INTENSITY, K (MPa√m)

Figure 5. In a), limited crack growth and significant crack blunting in hydrated dentin is seen. Crack growth data obtained for cortical bone specimens is shown in b).

stress functions deduced using Eq. 2, with bridging zone lengths of $L \sim 5 - 5.8$ mm.

Time-dependent cracking/blunting results

In hydrated dentin, crack growth was limited and only occurred within the first hour, after which the load continued to drop, but microscopic observations revealed this was due to creep deformation/crack blunting (Fig. 5a). In cortical bone, however, crack growth readily occurred at growth rates above 10^{-9} m/s, and a plot of crack velocity, da/dt, versus applied stress intensity, K, is shown in Fig. 5b. Time-dependent crack blunting occurred in cortical bone as well, but over much longer time scales. Indeed, growth could not be measured below 10^{-9} m/s because crack blunting essentially caused crack arrest over the time scales (days) needed to collect such data.

DISCUSSION

Rising R-curve behavior was observed in both dentin and cortical bone (Fig. 1), with crack bridging identified as the primary toughening mechanism responsible in all cases (Figs. 2 - 4). The present R-curve results for human bone are similar to those reported in the only known independent study.[16] For hydrated dentin, a steady-state toughness is quickly reached, which appears to be the result of extensive crack blunting (Fig. 5a) leading to larger crack-opening displacements than in dehydrated dentin or bone. For bridged cracks, the extent of the bridging zone behind the crack tip is generally thought to be controlled by a critical crack-opening displacement, after which the bridges fail and no longer provide toughening. Due to the larger crack openings associated with hydrated dentin, it appears that this critical crack-opening displacement is achieved after less crack extension than with dehydrated dentin, which did not show such blunting behavior. Furthermore, although cortical bone demonstrated crack blunting behavior similar to hydrated dentin, this occurred over time scales much longer (several days) than required for R-curve measurements, and accordingly R-curves in cortical bone samples also did not show a plateau.

The bridging stress results in Fig. 4 appear to be fairly consistent with the only previous related study by Pezzotti and Sakakura,[17] who used Raman spectroscopy to quantify the bridging stresses in bovine bone. Although they measured local bridging stresses as high as 300 MPa within ~10 μm of the crack tip, the stresses fell off significantly to ~10 – 50 MPa at distances of ~100 μm from the crack tip. Such values are in line with the present results, which indicate a average bridging stresses of ~7 – 17 MPa at the same position, especially considering that bovine bone is typically found to have higher toughness than human bone.[18,19]

Results in Fig. 5b indicate that time-dependent crack growth can occur in bone under sustained (non-cyclic) *in vitro* loads at stress intensities lower than the nominal crack-initiation toughness, K_o, measured on the R-curve (Fig. 1a). This behavior is similar to many ceramics, and implies that bone can fail subcritically at stresses well below those required for overload (instantaneous) fracture. Indeed, measurable crack growth was observed at stress intensities as low as 1.33 MPa√m, or 35% lower than K_o (Fig. 5b). Although the mechanism of such cracking currently is not clear, it does imply that the R-curve alone is not sufficient to fully characterize the fracture behavior of cortical bone; some consideration must be made for its time-dependent crack-growth properties.

Furthermore, both dentin and cortical bone appear to demonstrate two regimes of time-dependant behavior, with an initial stable crack-growth period during the first few minutes for dentin and the first several hours for bone. This is followed by the suppression of cracking due to significant (time-dependent) crack-tip blunting. In dentin, crack blunting occurs quite rapidly, causing crack arrest within the first hour (Fig. 5a), which allows very little crack extension to occur in the hydrated state under constant displacement conditions. In contrast, crack blunting occurs much more slowly in hydrated cortical bone, allowing time-dependent cracking to occur more readily (Fig. 5b). In this case, it is only at very slow growth rates (e.g., < 10^{-9} m/s) that there is adequate time for significant crack blunting to suppress cracking.

These observations imply that when mineralized tissues, such as bone and dentin, are subjected to sustained loading, there is a mutual competition between time-dependent crack blunting and subcritical crack extension. At higher stress intensities, the rate of forward progress of the crack tip exceeds the blunting-induced lateral displacement of the crack sides, such that considerable crack extension may be achieved, as seen for bone in Fig. 5b. However, at lower driving forces/growth rates, crack blunting becomes the more prominent mechanism, and although the load continues to drop during the experiment, this is primarily due to creep/blunting with very little concomitant crack extension. Thus, even without remodeling, dentin and bone have a mechanism to arrest subcritical cracks; this is especially important for dentin which essentially does not remodel *in vivo*. Furthermore, differences in the time dependence in bone and dentin for this creep/blunting behavior may shed light on the nature of the intrinsic fracture mechanisms, which are still uncertain. Indeed, since cortical bone and dentin essentially represent different microstructural arrangements of apatite impregnated collagen fibers, determining how fracture processes differ in these two materials should help elucidate the micro-mechanisms responsible for fracture, and how these mechanisms are affected by microstructural factors. Since cracks in bone tend to deflect along the boundaries of osteons, or cement lines, it may be the local properties of these weak microstructural paths which dictate the differences in crack blunting behavior.

CONCLUSIONS

Based on a study of the fracture toughness behavior of human cortical bone and elephant dentin, the following conclusions can be made:

1. Rising R-curve behavior was observed in both cortical bone and dentin, with such behavior attributed to the formation of uncracked ligament bridges in the crack wake, as identified by microscopy and X-ray tomography.

2. The bridging stress distributions in cortical bone were deduced using multi-cutting compliance experiments, with peak stresses of $\sim 7 - 17$ MPa found near the crack tip.

3. In the hydrated state, dentin demonstrated considerable time-dependent crack blunting due to creep deformation, leading to larger crack openings, premature destruction of bridges, and plateaus on the R-curves. In dehydrated dentin, no such blunting was observed, and R-curves rose linearly until the completion of the test (i.e., to $\Delta a \sim 5$ mm) due to the smaller crack openings.

4. Cracks in hydrated human cortical bone blunted as well, but much more slowly than in dentin, with similar degrees of blunting seen over several days instead of hours. Accordingly, the R-curves did not reach plateau values, and subcritical crack growth was observed for growth rates above 10^{-9} m/s. At slower growth rates, crack blunting behavior dominated and arrested cracks.

5. The fact that cracks in cortical bone undergo blunting over time scales that are much longer than in dentin, despite the fact that both are composed of collagen fibers reinforced with nanocrystalline apatite, is thought to be related to differences in their microstructures at coarser dimensions. Indeed, cracks in bone tend to deflect around the osteons and propagate preferentially along the weaker cement lines, which presumably act to keep the crack tip sharp.

ACKNOWLEDGEMENTS

This work was supported by the National Institutes of Health under Grant No. 5R01 DE015633 (for RKN) and by the Director, Office of Science, Office of Basic Energy Science, Division of Materials Sciences and Engineering, Department of Energy under No. DE-AC03-76SF00098 (for JJK and ROR). We also wish to thank Dr. A. P. Tomsia, Lawrence Berkeley National Laboratory, Berkeley, CA for his continued support, Drs. C. Puttlitz and Z. Xu, San Francisco General Hospital, San Francisco, CA for supply of the human cortical bone, and Ms. C. Kinzley, Curator, Oakland Zoo, Oakland, CA for supplying the elephant dentin.

REFERENCES

[1] J. H. Kinney, S. J. Marshall and G. W. Marshall, "The mechanical properties of human dentin: A critical review and reevaluation of the dental literature," *Crit. Rev. Oral Biol. Med.*, **14** [1] 13-29 (2003).

[2] J.-Y. Rho, L. Kuhn-Spearing and P. Zioupos, "Mechanical properties and the hierarchical structure of bone," *Med. Eng. Phys.*, **20** 92-102 (1998).

[3]"ASTM E399-90 (Reapproved 1997)"; in *Annual Book of ASTM Standards, Vol. 03.01: Metals- Mechanical Testing; Elevated and Low-temperature Tests; Metallography.* ASTM, West Conshohocken, Pennsylvania, USA, 2002.

[4]J. J. Kruzic, R. K. Nalla, J. H. Kinney and R. O. Ritchie, "Crack blunting, crack bridging and resistance-curve fracture mechanics in dentin: Effect of hydration," *Biomaterials,* **24** [28] 5209-21 (2003).

[5]R. K. Nalla, J. J. Kruzic, J. H. Kinney and R. O. Ritchie, "Mechanistic aspects of fracture and R-curve behavior of human cortical bone," *Biomaterials,* in press (2004).

[6]A. Saxena and S. J. Hudak Jr., "Review and extension of compliance information for common crack growth specimens," *Int. J. Fract.,* **14** [5] 453-67 (1978).

[7]J. H. Kinney and M. C. Nichols, "X-Ray Tomographic Microscopy (XTM) Using Synchrotron Radiation," *Annu. Rev. Mater. Sci.,* **22** 121-52 (1992).

[8]F. H. Wittmann and X. Hu, "Fracture process zone in cementitious materials," *Int. J. Fract.,* **51** [1] 3-18 (1991).

[9]R. Ballarini, S. P. Shah and L. M. Keer, "Crack growth in cement-based composites," *Eng. Fract. Mech.,* **20** [3] 433-45 (1984).

[10]M. Wecharatana and S. P. Shah, "A model for predicting fracture resistance in reinforced concrete," *Cem. Concr. Res.,* **13** 819-29 (1983).

[11]X.-Z. Hu, E. H. Lutz and M. V. Swain, "Crack-tip-bridging stresses in ceramic materials," *J. Am. Ceram. Soc.,* **74** [8] 1828-32 (1991).

[12]R. M. L. Foote, Y.-M. Mai and B. Cotterell, "Crack growth resistance curves in strain-softening materials," *J. Mech. Phys. Sol.,* **34** [6] 593-607 (1986).

[13]Y.-W. Mai and B. R. Lawn, "Crack-interface grain bridging as a fracture resistance mechanism in ceramics: II, Theoretical fracture mechanics model," *J. Am. Ceram. Soc.,* **70** [4] 289-94 (1987).

[14]H. F. Bueckner, "A novel principle for the computation of stress intensity factors," *Z. Angew. Math. Mech.,* **50** [9] 529-46 (1970).

[15]T. Fett and D. Munz, *Stress Intensity Factors and Weight Functions.* Computational Mechanics Publications, Southampton, UK, (1997).

[16]D. Vashishth, J. C. Behiri and W. Bonfield, "Crack growth resistance in cortical bone: Concept of microcrack toughening," *J. Biomech.,* **30** [8] 763-69 (1997).

[17]G. Pezzotti and S. Sakakura, "Study of the toughening mechanisms in bone and biomimetic hydroxyapatite materials using Raman microprobe spectroscopy," *J. Biomed. Mater. Res.,* **65A** [2] 229-36 (2003).

[18]J. D. Currey, "Mechanical properties of vertebrate hard tissues," *Proc. Instn. Mech. Engrs.,* **212H** 399-412 (1998).

[19]T. L. Norman, D. Vashishth and D. B. Burr, "Fracture toughness of human bone under tension," *J. Biomech.,* **28** [3] 309-20 (1995).

THE EFFECT OF ZIRCONIA ADDITION ON GLASS PENETRATION RATE AND MECHANICAL PROPERTIES OF CERAMIC-GLASS COMPOSITES

Deuk Yong Lee
Daelim College of Technology
526-7, Bisan-Dong
Anyang 431-715, Korea

Se-Jong Lee
Kyungsung University
110-1, Daejeon-Dong
Busan 608-736, Korea

Il-Seok Park
Yonsei University
134, Shincheon-Dong
Seoul 120-749, Korea

Bae-Yeon Kim
University of Incheon
177, Dohwa-Dong
Incheon 402-749, Korea

ABSTRACT

Alumina/zirconia and spinel/zirconia-glass dental crown composites were prepared by die-pressing and melt infiltration to investigate the effect of zirconia addition on glass infiltration rate and mechanical properties of the composites. The infiltrated distance was parabolic in time described by the Washburn equation and the penetration rate decreased due to the reduction in pore size as the zirconia content rose. The optimum strength of the alumina/zirconia and spinel/zirconia-glass composites was observed when 15 wt% and 20 wt% of zirconia were added, respectively. The optimized strength of the alumina/zirconia and spinel/zirconia-glass composites was 525 MPa and 383 MPa, respectively, which is highly applicable to dental crown materials

INTRODUCTION

All-ceramic dental crowns consisting of glasses and ceramics have been considered as the method of choice for esthetic restorative treatment in dentistry because of hardness, wear resistance, chemical inertness, nontoxicity and strength.[1-5] Among them In-Ceram crown that is manufactured by presintering of alumina cores formed by slip casting and subsequent infiltration of glass into the porous cores possesses a superior flexural strength and reasonably high fracture toughness.[1,2] The use of melt-infiltration in all ceramic dental crowns provides near-net shape forming process (NNS) having low shrinkage for accurate fit, which is prerequisite for biomechanical components and dental crowns. NNS includes sintering of alumina at 1120°C to develop a skeleton of fused alumina particles and a subsequent infiltration of the porous structure with lanthanum aluminosilicate glass at 1100°C for the densification.[1-5] Recently, easy shaping was accomplished by using CAD/CAM process.[6] This process may allow the economical rapid prototyping of complex three-dimensional shaped objects like dental bridges.

Lee et al.[3,7] reported that spinel calcined for 1h at 1200°C and alumina having particle size of 2.85 μm were highly effective to mechanical properties of the glass-ceramic dental composites. In the present study, zirconia[8] having a higher fracture toughness of 11 MPam$^{1/2}$ was added into spinel and alumina to improve the fracture toughness of the ceramic-glass composites because zirconia acted as an inclusion for spinel or alumina particles. La_2O_3-Al_2O_3-SiO_2 glass[2-5,7,9] was infiltrated into porous spinel/zirconia and alumina/zirconia performs to obtain the densified the dental ceramic composites and then the influence of zirconia addition on glass penetration rate and mechanical properties of the composites is examined.

EXPERIMENTAL PROCEDURE

Powder preparation procedure of zirconia having a composition of 3mol% Y_2O_3-1.5mol% Nb_2O_5-95.5mol% ZrO_2 and spinel (0.94 μm, Sumitomo Co., Japan) were described elsewhere.[8] Alumina powder (AL-M43, 99.9%, Sumitomo Chemical Co., Japan)/zirconia and spinel/zirconia were prepared by adding zirconia into alumina and spinel in 5 wt% intervals in the range of 0 to 20%(40%) using ball milling for 24 h. The milled slurries were dried, sieved through a 100-mesh screen, die-pressed into disks and then isostatically pressed at 140 MPa. The green compacts were presintered for 2 h at 1120 °C[3-5,9] or 1100°C[7] in air, respectively. Disks having a dimension of 20 mm in diameter and 1.7 mm in thickness were used for mechanical property measurements, 10 mm and 20 mm for glass penetration behavior.

The La_2O_3-Al_2O_3-SiO_2 glass[7,10] was prepared by melting the desired oxides in a platinum crucible at 1400°C, quenching the melt in water, and then grinding into a powder using a disc mill (Pulverisett 13, Fritsch GmbH, Germany). The glass powder-water slurry was placed on the partially sintered alumina/zirconia and spinel/zirconia preforms and infiltrated for up to 4 h at 1100°C (1080°C) with a rate of 30°Cmin^{-1} and then furnace cooled. The disk-type composites were polished to a 1 μm finish. The final thickness of the composites was 1.7 mm. For the measurements of the penetration behavior, the glass was infiltrated into the performs at 1100°C and 1080°C for 0.1 h to 2 h in the interval of 10 min and sectioned along the diameter using a diamond saw and then penetration distance of the composites was determined by an optical microscopy (Kanscope, Sometech, Korea).

The strength of the composites was evaluated by a flat-on-three-ball biaxial flexure testing. The specimens were broken using a biaxial strength fixture at a stress rate of 23 MPas^{-1}.[2,3,7] A Vickers indentation of 196N was placed on the center of the tensile faces of the five test specimens to measure fracture toughness. A drop of silicon oil was applied to minimize moisture assisted subcritical crack growth. For the calculation of the fracture toughness, the hardness to modulus ratio was estimated from the measurements of the diagonal dimension of Knoop indentation.[2,3)]

RESULTS AND DISCUSSION

Particle size distribution of ceramic powders is shown in Fig. 1. The mean particle size of alumina, spinel and zirconia is 2.85 μm, 3.0 μm and 0.49 μm, respectively, implying that fine zirconia particles may fill the voids of skeletal structure formed by coarse alumina and spinel particles. However, relative density remained almost constant as the zirconia content rose, whereas, shrinkage of alumina/zirconia and spinel/zirconia decreased and remained constant, respectively, as zirconia was added up to 20%, indicating that zirconia may retard the densification of the alumina and the spinel particles. Zirconia may hinder the grain growth of alumina and spinel because of the zirconia migration into grain boundaries,[10)] resulting in smaller pore size as depicted in Fig. 2 (less than 0.4 μm).

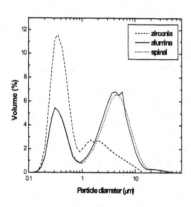
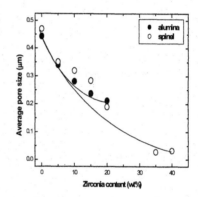

Fig. 1. Comparison of particle size distribution.

Fig. 2. Pore size variation as a function of zirconia content.

Strength and fracture toughness of alumina/zirconia and spinel/zirconia-glass composites are shown in Fig. 3. The optimum mechanical properties were observed for the composites containing 15 wt% zirconia (Fig. 3(a)) and 20 wt% zirconia (Fig. 3(b)), respectively. Guazzato *et al.*[11)] reported that the greater strength and toughness of the glass-infiltrated alumina/zirconia and spinel/zirconia over the glass-infiltrated alumina and spinel may be due to the synergic combination of the phase transformation of the zirconia grains and the crack deflection related to the alumina and spinel grains. The crack was deflected by the high strength alumina and spinel grains, whereas it propagated through the transformed zirconia grains as shown in Fig. 4. Phase transformation of highly transformable zirconia[8] occurred at the crack tip caused compressive stresses, which was effective to crack shielding. Such benefits were likely to be counteracted by the presence of the greater porosity. Strength decrement when 20 or 25 wt% of zirconia were added is due probably to the difficulty in densification of the composites caused by the smaller pore size as demonstrated in Fig. 2. The infusion of liquid glass into the gap between the skeleton of fused alumina/zirconia and spinel/zirconia particles may be hampered.

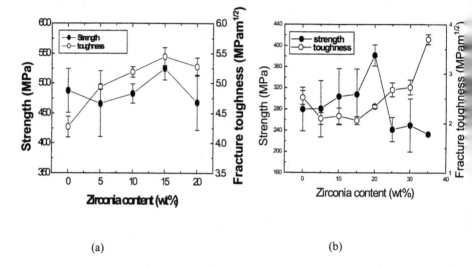

(a)   (b)

Fig. 3. Flexural strength and fracture toughness of the alumina/zirconia and spinel/zirconia-glass composites.

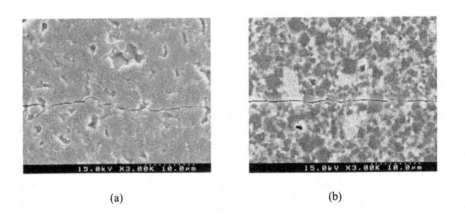

(a)   (b)

Fig. 4. SEM backscattered electron images of the indented alumina/15% zirconia and spinel/20% zirconia-glass composites.

The time to achieve a given penetration depth is believed to be directly proportional to the pore radius.[12] Therefore, the extent of infiltration of the glass may be a function of pore size because infiltration was driven by capillarity.[12] Fig. 5 indicated that the infiltration distance was parabolic in time as described by the Washburn equation and the penetration rate constant, K, increased with decreasing the zirconia content due to the increase in pore size (Fig. 2). Especially, the infiltration rate of both alumina/zirconia and spinel/zirconia preforms containing zirconia was retarded significantly as expected due to the smaller pore size.

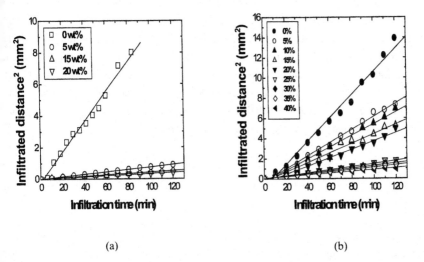

(a)   (b)

Fig. 5. (Infiltrated distance)^2 vs. infiltrated time for the (a) alumina/zirconia and (b) spinel/zirconia preforms having different zirconia content.

Lee et al.[3] noted that excellent mechanical properties of the alumina-glass composites can be achieved when average alumina particle size (bimodal distribution) and pore size were about 3.0 μm and 0.44 μm, respectively. It was found that smaller pore size less than 0.1 μm and monosize particle distribution were deleterious to densification and strength. Also, submicrometer grains in the partially sintered alumina may dissolve in the glass and thus infusion of liquid glass may be hampered. In the present study, pore size of the alumina/20wt% zirconia and spinel/40wt% zirconia preforms is 0.2 μm and 0.03 μm, respectively, which is detrimental to glass penetration, resulting in poor densification and lower mechanical properties.

Strength of alumina and spinel-glass composites rose from 488 MPa and 280 MPa to 525 MPa and 382 MPa when zirconia was added by 15 wt% and 20 wt%, respectively, due to the synergic effects of the phase transformation of the zirconia grains and the crack deflection related alumina and spinel grains. However, the glass penetration time decreased due to the reduction in pore size as the zirconia content was raised, resulting in longer glass infiltration time for the composites.

CONCLUSIONS

Alumina/zirconia and spinel/zirconia-glass dental composites were prepared by die-pressing and melt infiltration to investigate the effect of zirconia addition on mechanical properties and glass infiltration rate of the composites. The infiltrated distance was parabolic in time described by the Washburn equation and the penetration rate decreased due to the reduction in pore size as the zirconia content rose. The optimum strength of the alumina/zirconia and spinel/zirconia-glass composites was observed when 15 wt% and 20 wt% of zirconia were added. The optimized strength of the alumina/zirconia and spinel/zirconia-glass composites was 525 MPa and 383 MPa, respectively, which is highly applicable to dental crown materials.

ACKNOWLEDGEMENTS

This work was supported by a research fund from the Daelim College of Technology.

REFERENCES

[1]W.D. Wolf, L.F. Francis, C-P. Lin and W.H. Douglas, "Melt-infiltration Processing and Fracture Toughness of Alumina-glass Dental Composites," *Journal of the American Ceramic Society*, **76** [10] 2691-94 (1993).

[2]D-J. Kim, M-H. Lee, D.Y. Lee and J-S. Han, "A Comparison of Mechanical Properties of All-ceramic Alumina Dental Crowns Prepared from Aqueous- and Non-aqueous-based Tape Casting," *Journal of Biomedical Materials Research*, **53** [4] 314-19 (2000).

[3]D.Y. Lee, D-J. Kim, B-Y. Kim and Y-S. Song, "Effect of Alumina Particle Size and Distribution on Infiltration Rate and Fracture Toughness of Alumina-glass Composites Prepared by Melt-infiltration," *Materials Science & Engineering A*, **341** [1-2] 98-105 (2003).

[4]D.Y. Lee, S-J. Lee, I-S. Park, J-W. Jang and B-S. Kim, "Glass-alumina Composites Prepared by Melt-infiltration: III. *In-vitro* Hertzian Contact Fatigue," *Journal of Korean Ceramic Society*, **40** [1] 662-66 (2003).

[5]S-J. Lee, "Wear Behavior of Alumina-glass Composites Prepared by Melt Infiltration," *Journal of Korean Ceramic Society*, **40** [9] 881-85 (2003).

[6]H. Hornberger, "Strength Microstructure Relationsips in a Dental Alumina Glass Composite," Ph.D. Dissertation, University of Birmingham, 1995.

[7]D.Y. Lee, D-J. Kim and Y-S. Song, "Properties of Glass-spinel Composites Prepared by Melt Iinfiltration," *Journal of Materials Science Letters*, **21** [15] 1223-26 (2002).

[8]D.Y. Lee, D-J. Kim and B-Y. Kim, "Influence of Alumina Particle Size on Fracture Toughness of (Y,Nb)-TZP/Al$_2$O$_3$ Composites," *Journal of the European Ceramic Society*, **22** [13] 2172-79 (2002).

[9]J-W. Jang, B-S. Kim, H-K. Kim and D.Y. Lee, "Correlation Between Thermal Expansion Coefficients of La$_2$O$_3$-Al$_2$O$_3$-SiO$_2$ Glasses and Strength of the Glass Infiltrated Alumina for All Ceramic Crown," *Materials Science Forum*, **449-452** 1193-96 (2004).

[10]S. Taruta, K. Kawashima, K. Kitajima, N. Takusagawa, K. Okada and N. Otsuka, "Influence of Zirconia Addition on the Sintering Behavior of Bimodal Size Distributed Alumina Powder Mixtures," *Journal of Japanese Ceramic Sociey*, **102** [2] 139-44 (1994).

[11]M. Guazzato, M. Albakry, M.V. Swain and S.P. Ringer, "Microstructure of Alumina- and Alumina/zirconia-glass Infiltrated Dental Ceramics," *Key Engineering Materials*, **240-242** 879-82 (2003).

[12]W.B. Hillig, "Melt Infiltration Approach to Ceramic Matrix Composites," *Communication of American Ceramic Society*, **71** [2] C96-99 (1988).

QUANTIFYING THE EFFECT OF DENTAL CERAMIC COLOR VARIABILITY ON RESTORATION COLOR VARIATIONS.

Veeraraghavan (V). Sundar and Julie Noriega
Dentsply Prosthetics, Ceramco Division
6 Terri Ln., Suite 100, Burlington, NJ 08016

Claire Allison Rutiser
CRMA, 102 Cameron Rd.,
Willow Grove, PA 19090

ABSTRACT

A sustained interest in better esthetics has gone hand in hand with better function and reliability of dental restorations. Research into the fidelity of shade reproduction has identified many factors that affect the color variability of a given restoration. These include, but are not limited to technique variability, difference in color perception among observers, process variations, and finally, variability in the color of dental materials. The goal of this article is to quantify the contribution of variability of different lots of the same shade of dental ceramics to the color variability of a restoration. Multiple lots of two common shades of opaque and translucent components of a conventional high-fusing and a low-fusing low-wear porcelain were evaluated using a spectrophotometer. The variability in terms of CIELAB color difference DE was found to be < 1.5 on an average, indicating a relatively small contribution from lot to lot variability. The variability in opaque porcelain was found to be less than that of translucent porcelain. There was no statistically significant difference between the variability of the high-fusing and that of the low-fusing porcelains.

INTRODUCTION

Many factors influence the final shade match of a dental restoration to the natural teeth surrounding it (Table 1). These factors may range from the purely psychological (e.g. patient preferences) through psychosomatic (e.g. color acuity and shade matching of dental office and lab personnel), mechanical (e.g. lighting conditions in dental office and dental lab) to the material (e.g. shade stability and fidelity of reproduction of restorative materials). Given the predominantly referential nature of most shade matches, there is definitely a subjective component to shade matching. The addition of descriptive communications such as shade maps or photographs and quantitative information such as colorimetric readings can reduce the subjective nature of the shade matching process.

Several recent studies have examined the quantitative contributions of the various steps of the restoration process. The most commonly used quantitative measures of color differences are CIELAB (Commission Internationale De L'éclairage, Vienna, Austria), and CMC (Color Measurement Committee, Society of Dyers and Colourists, Great Britain) units [1].

The visual acuity of dental office personnel was compared to that of the general population in a study at the US Naval Dental Research Institute [2]. This study suggests that patients, who may not be as trained I discriminating small color differences associated with tooth color, were not as discriminating in their ability to identify color differences between composite restorations and the tooth as were dental professionals. The general population had a median color discriminating ability at a CMC DE of 2.29. In contrast, dental office professionals were able to discriminate median color differences of CMC DE of 1.78.

In a related study, A CMC color difference 1.78 or lower (CIELAB DE < 2.5) was considered to be a definitive match, and one of 2.69 (CIELAB DE < 4.0) was considered to be an approximate match by correlating CMC, CIELAB and visual assessments of shade matching. [3]. As a result these and other studies, a CIELAB difference of DE = 2.5 or lower is generally accepted to be the limit of quantitative color difference for an acceptable color match.

Table 1: Factors that could affect color difference perceptions in dental restorations.

Materials Manufacturer	Dental Office	Dental Laboratory	Patient
<ul><li>Manufacturing color accuracy and lot to lot variability</li><li>Shade stability</li><li>Effect of storage</li><li>Effect of oral environments</li><li>Technique sensitivity of materials.</li></ul>	<ul><li>Color acuity of personnel</li><li>Lighting conditions</li><li>Color reference used (shade guide system)</li><li>Color communication system used.</li></ul>	<ul><li>Color acuity of personnel</li><li>Lighting conditions</li><li>Processing</li><li>Materials system effects</li><li>Color reference used (shade guide system)</li><li>Color communication system used.</li></ul>	<ul><li>Color acuity</li><li>Esthetic preferences</li></ul>

Several studies that compared visual shade matching to instrumental shade matching have concluded that shade determination by visual means was somewhat inconsistent, and almost always less accurate than instrument assisted shade determination [4-7].

The cumulative effect of the various factors is particularly reflected in the variability of dental restoration colors as reproduced by commercial laboratories. In North America, it is fairly common for laboratories to create restorations based on written prescriptions alone, and this could have a further dilutive effect on the accuracy of shade reproduction. A recent study examined the variability in color reproduction for metal ceramic crowns fabricated by commercial dental laboratory technicians. They determined the average color difference variance among labs was approximately 6 CIELAB DE units for both middle and incisal thirds of a crown, thereby jeopardizing the clinical acceptability of shade matching. Mean color differences from shade tabs for individual labs ranged from 3.5 to 11.1 CIELAB DE units in this study [8].

One factor that has not been comprehensively quantified in this matrix is the lot to lot variability of the color of dental ceramics. Such a variation could be a contributing factor to the variability in the output of commercial dental labs as well as in the color match to a restoration.

One particular study that has explored this area was targeted at evaluating an automated color system, and analyzed all-ceramic (medium pressure injection moldable leucite-reinforced glass-ceramic) materials [9]. In analyzing their data, the authors concluded that firing procedure differences led to on a small (CIELAB DE < 2.2) difference in the color of restorative materials. They recommended the use of spectrophotometric devices as a means of quality control in order to decrease color tolerances between different batches of the same material. However, most dental ceramic manufacturers still use trained visual observers to perform color quality assurance procedures on their products, and appear to have achieved acceptable color tolerances.

The purpose of this study, therefore, was to evaluate the lot-to-lot variability of commercially available dental ceramics. We specifically set out to design a retrospective evaluation where materials deemed to be commercially acceptable could be analyzed using spectrophotometry, and the visual acuity of trained color QA personnel could be quantified. This can help to evaluate the contribution of color variability in commercial dental ceramics to the variation in results achieved with restorations. It can also help to evaluate the relative importance of various color attributes (hue, value, chroma etc.) in relative order of importance.

MATERIALS AND METHODS

Multiple lots of two common shades (Shades A2 and A3, Vita Lumin Shade Guide, Vita Zahnfabrik H. Rauter GmbH and Co. KG, Bad Säckingen, Germany) were chosen for evaluation. These two shades can comprise up to 40% of all prescribed shades. These lots were all commercially distributed materials, that had been put through a visual quality assurance color matching regime, and deemed to be an acceptable match to specific lots of standard materials. Both opaque and translucent dentin components of two systems were chosen. A conventional high-fusing (Ceramco II Porcelain System, Dentsply Ceramco, Burlington, NJ) was used as a primary evaluation material for both translucent and opaque ceramics. Two dentin shades (A2 and A3) of a low-fusing low-wear (Finesse® Low Fusing Porcelain System, Dentsply Ceramco, Burlington NJ) porcelain were also evaluated to include any differences introduced by ceramic firing characteristics.

Powdered samples of a uniform weight (2.5g each) were used to eliminate variations in thickness of the disks measured. This amount was determined to yield an acceptable approximation of infinite optical thickness of these materials. At the infinite optical thickness, the background of color comparison has no effect on the color observed [10].

The standardized amount of powder was placed in a custom fabricated three-part uniaxial metal die. The die was then sealed and placed between the plates of a hydraulic press (Carver Hydraulic Unit Model 3912, Fred S. Carver Inc., Menomenee Falls, WI). An applied load of 2 metric tons was used to compact the powder. The compacted powder disk was removed from the die, and placed on a cordierite honeycomb sagger tray (Phoenix Sagger Trays, Dentsply Ceramco, Burlington, NJ) for firing. A dental porcelain furnace (Phoenix Quick Cool, Dentsply Ceramco, Burlington, NJ) was used to sinter the disks. Standardized sintering programs were used for opaque and dentin materials. Typically, these followed the recommended sintering profiles for the porcelain materials in a crown form. The additional mass of the disks was compensated for using an increase in high temperature (~40°C), and an increase in hold time (~1min). Uniformly sintered buttons resulted from this procedure.

As part of the visual acceptance procedure, trained evaluators attempted to detect a difference between pairs of these disks as manufactured (the batch) and pairs of disks fired from a designated lot of standard material of that shade (the standard). An acceptable color match was deemed to have been achieved when the trained evaluators were no longer able to distinguish the standard pair and the batch pair in a blind test. All materials used in these tests had been deemed to be acceptable using this visual evaluation procedure. All evaluators had a minimum of one year's experience in dental ceramic color quality assurance. Their visual acuity had been evaluated using the Farnsworth-Munsell 100-Hue Test for Color Vision (Munsell Color, Baltimore, MD), and found to be significantly better than average.

The disks of various materials were then evaluated using a spectrophotometric colorimeter. The instrument used was a SpectraFlash SF500 integrating sphere colorimeter (DataColor, Lawrenceville, NJ). The instrument is a dual-beam spectrophotometer, using diffuse illumination and an 8° viewing measurement geometry. A xenon pulse lamp is used to simulate the CIE D65° measurement condition, which was used in all these measurements. A wavelength range of 400-700nm was used, and the reflectance output was analyzed using a 152mm diameter integrating sphere and a proprietary analyzer with a dual 256 diode array. The proprietary ChromaCalc2.0 software was used by the computer controlling the instrument to convert output into CIELAB units. This colorimeter was calibrated using a reference white tile provided by the manufacturer and a high performance black trap prior to each measurement series (Fig. 1).

Fig.1: Calibration range of SF600, reflectance (%) vs. wavelength (nm)

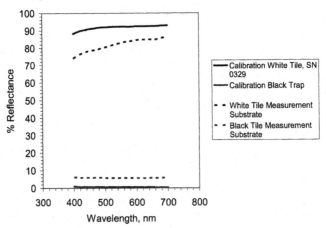

Disks were measured using a small aperture view (SAV) port, 9mm in diameter. A 5mm area was used for the measurement. Two measurements were carried out for each disk, against standardized black and white backgrounds. The instrument then calculated the intrinsic color of the disk at infinite optical thickness, as a function of the two reflectance curves obtained (Fig. 2).

An output in CIELAB units, as well as a contrast ratio measurement of opacity (on a scale of 0-100, with 0 indicating perfect transparency) was the result of this measurement series. A color difference from a standard disk was also calculated for each lot.The following instrument error evaluations were performed for this study:

- Instrument calibration against master instruments using BCRA standard color tiles.
- Instrument repeatability evaluation using a calibration white tile.

Twenty lots (batches) of each shade of opaque and dentin material, in high- and low-fusing porcelains were chosen, manufactured over a period of 1998-2000. A common standard was used for each of these lots. For each material, the characteristics evaluated were:

- Lot to lot variation in batch materials.
- Color difference between standards and batches in each manufacturing lot.
- Color variation in standard materials, caused by slightly different firing conditions.

These measurements taken together could then provide a comprehensive picture of the error contributions related to equipment limitations, as well as visual color matching methodology.

The average and standard deviation for each component (CIELAB L*, C* and h*) was also evaluated independently, to analyze their individual contributions.

Fig.2: Reflectance (%) vs. wavelength (nm) for an A1 shade of low-fusing dental ceramic

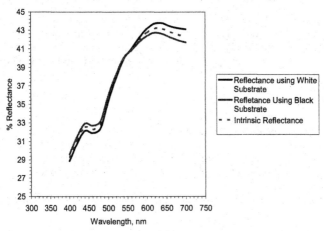

RESULTS AND DISCUSSION

Instrument Error Quantification
    Calibration: The Spectraflash 500 instrument was calibrated against a bank of master instruments, over a range of hue, chroma and value, using twelve BCRA ceramic tiles as color

Fig.3: Calibration error of the SF500 measured against a master instrument. The X axis shows the color numbers listed above for BCRA tiles.

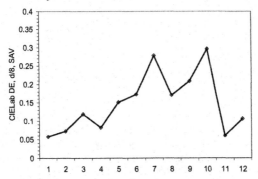

standards. The average variability in color was determined to be a DE < 0.15 CIELAB units over the range of color, with a maximum of 0.30 DE units for any given tile. This was a relatively small contribution to the overall error. The color range used was pale gray (1), mid gray (2), deep gray (3), deep pink (4), red (5), orange (6), bright yellow (7), green (8), cyan (9), deep blue (10), white (11) and black (12). There was no consistent trend in color error with hue. The instrument is toleranced at DE < 0.4 (maximum), and DE < 0.25 (average), for this test (Fig. 3).

*Instrument Repeatability:*The repeatability of the instrument was evaluated by twenty measurements of a calibration white tile (#0329) provided by the manufacturer. Over a series of 25 measurements, the average CIELAB DE was found to be DE = 0.02, with a standard deviation of 0.02. This was also judged to be a relatively small contribution to the overall error of measurement.

Material Color Variation Evaluation

*Lot to lot Variation in Batch Materials:*The color variation among batches alone was initially evaluated, as this is the variation experienced by the end user of dental ceramics. The results for opaque and translucent shades of high and low fusing porcelains are presented below, in terms of averages and standard deviations of color differences from standards (Fig. 4).

Fig. 4: *Reflectance curves for LF (low fusing) and HF (high fusing) dentin standards.*

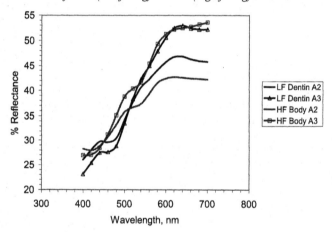

Results are expressed in terms of average ± standard deviation, in CIELAB and CIELCh units (Table 2).

Color Difference from Standard Materials: The color difference between the standard for each material and each batch sample was measured next. This would help to quantify the variability in the process, and identify any differences between materials of different translucencies and different firing properties. Again, the average difference in color is given here as a result, combined with the standard deviation (Table 3).

Table 2: Color variation between lots of opaque and translucent dental ceramics (average±standard deviation)

Material	Fusion	Shade	Value, L*	Chroma, C*	Hue, h*	Contrast Ratio
Opaque	High-Fusing	A2	74.8±0.3	16.8±0.1	77.7±0.1	100
	High-Fusing	A3	73.9±0.8	20.0±0.5	75.6±0.5	100
Dentin	High Fusing	A2	67.6±0.4	9.8±0.2	86.4±0.4	94.1±0.6
	High-Fusing	A3	66.6±.4	12.4±0.2	83.7±0.4	95.8±0.4
Dentin	Low-Fusing	A2	69.9±0.4	14.8±0.2	78.5±0.3	97.3±0.7
	Low-Fusing	A3	71.4±0.3	20.9±0.2	76.0±0.2	95.9±0.7

Even accounting for systematic errors of about 0.15 DE units from the calibration error, and 0.02 DE units from repeatability, the resultant average DE is consistently less than 1.0. This indicates a fairly tight tolerance. The limits in terms of one, two and three standard deviations (Σ) are given (Table 3a), assuming a normal distribution of errors.

Table 3: Color difference from a standard for dental ceramics CIELAB units.

Material	Fusion	Shade	DL*	DC*	DH*	DE*
Opaque	High-Fusing	A2	-0.01±0.37	0.05±0.25	0.01±0.28	0.33±0.28
	High-Fusing	A3	-0.44±0.41	0.19±0.40	0.04±0.35	0.86±0.58
Dentin	High Fusing	A2	0.74±0.50	-0.25±0.24	0.14±0.61	0.84±0.50
	High-Fusing	A3	-0.30±0.38	0.26±0.22	-0.34±0.32	0.50±0.32
Dentin	Low-Fusing	A2	-0.01±0.39	0.05±0.20	0.05±0.19	0.38±0.30
	Low-Fusing	A3	0.01±0.31	0.03±0.23	0.01±0.03	0.33±0.20

We may assume DE < 2.4 as the limit of detectable visual color difference for dental professionals. This number derives from the limit for a definitive match found in the study by Sakashita et al. (2002). This indicates that the color QA process used currently results in visually indistinguishable samples in over 99% of matches. The average DE detected here also significantly lower than the DE < 2.2 limits measured in the study by Rinke et al. (1996).

Table 3a: Range of Statistically Possible CIELAB DE values

Material	Fusion	Shade	-2Σ	-1Σ	DE*	+1Σ	+2Σ	+3Σ
Opaque	High-Fusing	A2	< 0	0.05	0.33	0.61	0.89	1.17
	High-Fusing	A3	< 0	0.28	0.86	1.44	2.02	2.60
Dentin	High Fusing	A2	< 0	0.34	0.84	1.34	1.84	2.34
	High-Fusing	A3	< 0	0.18	0.50	0.82	1.14	1.46
Dentin	Low-Fusing	A2	< 0	0.08	0.38	0.68	0.98	1.28
	Low-Fusing	A3	< 0	0.13	0.33	0.53	0.73	0.93

Time Trends in Lot to Lot Variation: The time trends in lot to lot variation were plotted from the earliest date of manufacture to the latest in each batch. No apparent time trend was discovered. Trends for the high-fusing translucent shades are shown below (Fig. 5)

Fig. 5: Time Trends in Color Variation for high fusing (HF) dentin ceramics.

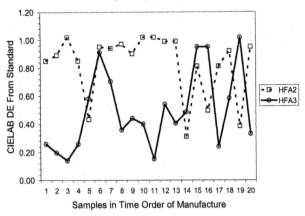

Samples in Time Order of Manufacture

Color Variation within Standard Buttons: The color variation within the various standard disks was evaluated next. A fresh pair of standard disks is fabricated for every shade match process. Variations caused by firing conditions, or possible aging of the standard materials could then lead to errors in color matching. The average of the color data of all standards was chosen as a virtual standard. Standard deviations in DL*, DC* and DH* color data are shown below as a measure of process variability. The average DE* color difference is also given (Table 4).

Table 4: Standard deviation of color differences between firings of standard disk, CIELAB units.

Material	Fusion	Shade	DL*	DC*	DH*	DE*
Opaque	High-Fusing	A2	0.33	0.28	0.27	0.33
	High-Fusing	A3	0.92	0.42	0.40	0.83
Dentin	High Fusing	A2	0.55	0.11	0.30	0.47
	High-Fusing	A3	0.61	0.16	0.38	0.51
Dentin	Low-Fusing	A2	0.61	0.52	0.61	0.71
	Low-Fusing	A3	0.72	0.13	0.19	0.61

Sensitivity Analysis: The ratio of standard deviation of a particular color variable to the average was calculated for CIELAB hue (h*), chroma (C*) and value (L*) for each material. All of the samples used in this analysis had been deemed to be acceptable visual color matches. A definite trend in range/average ratios may suggest that the trained human eye is more sensitive to a color attribute in relation to others. The results are tabulated below (Table 5).

These results suggest that the human eye allows a very small variation in hue. Value is the next smallest variation allowed. A relatively wider tolerance is maintained for chroma. This order of allowed variance is independent of opacity or translucency of the samples analyzed. It has been assumed in several discussions that the human eye is most sensitive to value in dental shade matching. This analysis would suggest that the trained observer is slightly more sensitive to hue than to value.

Table 5:C ovariance trends for approved batches of dental ceramics, in CIELAB units.

Material	Fusion	Shade	Value, L*	Chroma, C*	Hue, h*
Opaque	High-Fusing	A2	0.40%	0.60%	0.13%
	High-Fusing	A3	1.08%	2.50%	0.66%
Dentin	High Fusing	A2	0.60%	2.27%	0.52%
	High-Fusing	A3	0.53%	1.99%	0.53%
Dentin	Low-Fusing	A2	0.56%	1.32%	0.34%
	Low-Fusing	A3	0.44%	1.12%	0.31%

SUMMARY

Twenty lots each of opaque and translucent dental ceramics that were deemed to visually acceptable color matches were analyzed using spectrophotmetric colorimetry. The error contributions from instrument repeatability and calibration were quantified. The average color difference from a designated standard for both opaque and translucent samples was CIELAB DE < 1. This is a low variation compared to the visually detectable limit of DE > 2.5. This is also a relatively small contribution to the color variation of a final restoration, which has been reported to vary from 3.5 to 11.5 DE units, with a mean of 6.0. In analyzing the sensitivity of the trained visual observer, the results suggest that a very tight tolerance is maintained for hue, followed by slightly wider tolerances for value and chroma. Overall, the contribution of commercial dental ceramic color variability to the variability of a final restoration appears to be quite small.

REFERENCES

1. Billmeyer, F.W. Jr, and M. Saltzman, *Principles of Color Technology*, John Wiley & Sons; ASIN: 047103052X, 2nd edition, April 1981.
2. Ragain, JC Jr., and W.M. Johnston, "Minimum color differences for discriminating mismatch between, composite and tooth color", *J Esthet. Restor. Dent.*, v13(1), p41, 2001.
3. Sakashita, K., S. Nagai, K. Ishibashi, H. Yamamoto, and D. Nathanson, "Comparison of CIELAB And CMC Formulas For Color Difference", *J. Dent. Res.*, v81, pA-321, Abstr. #2545, 2002.
4. Okubo SR, A. Kanawati, M.W. Richards, and S. Childress, "Evaluation of visual and instrument shade matching", *J. Prosthet. Dent.*, v80 (6), p642.. 1998.
5. Paul, S., A. Peter, L. Rodoni, N. Pietrobon, and C.H.F. Hammerle, Clinical Comparison of a Spectrophotometric Shade Matching System with Conventional Shade Matching, J. Dent. Res., 2002; v81, pA-234, Abstr. #1783.
6. Nathanson D., S. Nagai, H. Yamamoto, T. Shioyama, and K. Ishibashi, "Evaluation of an automated color reproduction method for dental porcelain restorations", *J. Dent. Res.*, v81, pA-234, Abstr. #1782, 2002.
7. Faber, F.-J., I. Treunert and T. Kerschbaum, "Comparison of visual and instrumental color discrimination", *J. Dent. Res.*, v81, pA-66, Abstr. #0315, 2002.
8. Douglas, D., and J. D. Brewer, "Variability of Porcelain Color Reproduction by Commercial Laboratories", *J. Dent. Res.*, v81, pA-234, Abstr. #1784, 2002.
9. Rinke S., A. Hüls, M.J. Kettler, "Colorimetric analysis as a means of quality control for dental ceramic materials", *Eur. J. Prosthodont. Restor. Dent.*; v4(3), p105, 1996.
10. Miyagawa, Y., and J. M. Powers, "Prediction of color of an esthetic restorative material", *J. Dent. Res.*, v62(5), p581, 1983.

Synthesis and Characterization of
Phosphate-Based Bioceramics

COMPARISON BETWEEN PRIMARY AND CLONAL OBTEOBLAST CELLS FOR *IN VITRO* ATTACHMENT STUDIES TO HYDROXYAPATITE

M.A. Griffin, I.O. Smith and M.J. Baumann
Department of Chemical Engineering and Materials Science
Michigan State University
2125 Engineering Building
East Lansing, MI 48824-1226

ABSTRACT

Osteoblasts (OBs) are bone forming cells commonly used with bioceramic scaffolds such as hydroxyapatite (HA) in *in vitro* studies to assess the biocompatibility of the ceramic scaffold for *in vivo* use. While an OB cell line (MC3T3-E1) is typically used to assess the biocompatibility of the ceramic scaffolds, a question has arisen regarding as to whether the HA surface reacts the same to MC3T3-E1 OBs as it would to primary osteoblasts. Specifically, are the surface, or zeta (ζ) potentials of the MC3T3-E1 and primary OBs identical? To therefore assess their suitability in comparison to primary cells for attachment to HA, primary OBs were isolated from fetal rat calvaria and plated onto polystyrene (PS). These pOBs were then multiplied to a sufficient population (3×10^6 cells) necessary for ζ-potential analysis, suspended at 1% in physiologic saline (PS) and ζ-potential data collected over a biologically relevant pH range (pH 7.3-7.5). Comparisons between these ζ-potential values and those of the MC3T3-E1 OBs were found to be similar, with potentials between -8.5 and -26.8 mV for pOBs and between -29.9 and -52.4 mV for MC3T3-E1 OBs, over the pH range from 7.3 to 7.5. Since electrostatic potential is a good indicator of the propensity for attraction between HA and OBs, these findings show that for in vitro studies, the MC3T3-E1 clonal cell line is an appropriate substitute for primary cells.

INTRODUCTION

An important step in the ongoing process of developing an ideal bone tissue scaffold is the study of OB attachment and differentiation to calcium phosphate-based ceramic scaffolds. By seeding OBs (bone-forming cells) onto such a ceramic scaffold, indicators of bone formation such as cell attachment, differentiation and growth can be measured. The success of such a study depends on the ability of the ceramic scaffold, in this case HA, to first promote the attachment and differentiation of the OBs. Calcium phosphate (CaP) bioceramics are frequently used in bone tissue engineering because of their excellent biocompatibility and ability to integrate with bone. HA in particular closely mimics the composition of the mineral component of human bone. It is the combination of these properties which has led to HA being widely accepted as a bone tissue engineering implant material. Porous HA scaffolds encourage OB attachment and differentiation, more so than dense HA scaffolds.[1-4]

A question has recently been raised regarding the appropriateness of using clonal OBs as a substitute for primary OBs for *in* vitro bone tissue engineering studies onto CaP ceramic scaffolds. More specifically, the concern is whether the ζ-potential of the MC3T3-E1 OBs is similar to the ζ-potential of primary OBs. The use of the MC3T3-E1 mouse osteoblast clonal cell line is widely accepted in studies of CaP bioceramics for bone tissue engineered scaffolds because of their proven ability to form calcified bone tissue *in vitro*.[5-7]

Ionic activity between OBs and CaP powders may be assessed by calculating the ζ-potential from the measured acoustophoretic particle mobility as described by O'Brien and Oja.[8-10] ζ-

potential analysis has been shown to be an accurate indicator of CaP bioceramic bonding to bone.[11, 12] The ζ-potential is calculated from the acoustophoretic mobility with a correction for the particle inertia $G(\alpha)^{-1}$, in an alternating field. This correction reduces the velocity amplitude of the particle motion and was derived using the Helmholtz-Smoluchowski equation. Particles suspended in an electrolyte which undergo spontaneous ionization, produce an acoustic wave $(ESA(\omega))$ when exposed to an applied voltage potential, because of differences in density between the system components $(\Delta\rho)$. $ESA(\omega)$ is then used to calculate the particle mobility, using the following relationship developed by O'Brien (Eq. 1)

$$\mu_d(\omega) = ESA(\omega) / \phi\Delta\rho c \qquad (1)$$

where ϕ is the volume fraction of suspended solid and c is the velocity of sound within the system. The value for mobility is coupled with the Helmholz-Smoluchowski equation and used to determine the ζ-potential (Eq. 2). The Helmholz-Smoluchowski equation includes a correction for the particle inertia $G(\alpha)^{-1}$ (Eq. 3).

$$\zeta = (\mu\eta / \varepsilon_0\varepsilon_r)G(\alpha)^{-1} \qquad (2)$$

$$G(\alpha)^{-1} = \left[1 - \frac{i\alpha(3 + 2\Delta\rho / \rho)}{9\{1 + (1 - i)(\alpha / 2)^{1/2}\}}\right] \qquad (3)$$

The surface charge is not readily measurable because of continuous ion activity at the particle/fluid interface. Therefore, the surface potential is estimated at the double layer by the ζ-potential, which in this study, is measured using electrokinetic sonic amplitude (ESA) methods.

MATERIALS AND METHODS
HA scaffold fabrication
A commercial HA powder (particle size diameter of $0.85\mu m$)[1], of vendor-specified medical grade purity, was obtained and used as manufactured. Green disc specimens were fabricated by uniaxially pressing 1.5 ± 0.002 g of dry powder in a 32 mm diameter pellet die at 6.9 MPa for one minute using a standard laboratory press. Discs were sintered in air for four hours at either 1360°C in a high temperature sintering furnace (CM Rapid Temp Furnace) by heating to the sintering temperature at a rate of ~10°C/min. Following sintering, samples were cooled to room temperature by a controlled furnace cool at a rate of ~10°C/min.

pOB Isolation
Primary fetal rat osteoblasts (pOBs) were used in this study. Rats were chosen over mice because of their larger size which allows for more precise dissection and subsequent digestion. To isolate primary osteoblast cells, fetal rat calvaria were dissected and digested in a mixture of enzymes. Five pregnant Sprague-Dawley rats were euthanized and the fetal calvaria harvested. Calvaria were individually removed from the surrounding bony tissue (Figure 1) and placed in Minimum Essential Medium (MEM) supplemented with Earle's Salts (ES) with 10% Fetal Bovine Serum (FBS). Calvaria were then removed from the MEM solution for further dissection,

---

[1] Berkeley Advanced Biomaterials Inc., San Leandro, Ca., USA

thereby eliminating the immature bone tissue. Cuts were made around the major suture spaces indicated as bold black lines, Figure 1) leaving a wide margin of immature bone completely surrounding the suture spaces.[13] Pieces of bone removed from between the suture spaces (indicated as 1-4, Figure 1) were rinsed in Hanks Balanced Salt Solution and once again placed in MEM supplemented with ES and 10% FBS.

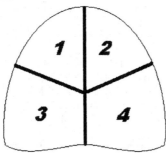

Figure 1 Schematic of fetal rat calvaria. Note dissected regions shown as 1-4. Dark lines indicate suture spaces.

An enzyme mixture consisting of 0.25% trypsin and 0.2% collagenase, mixed in a 1:1 ratio, was used for digestion. Approximately 1mL of the enzyme mixture was used for every four calvaria. In order to ensure thorough mixing during digestion, calvaria were placed in a sterile conical tube with the enzyme mixture and shaken. The tube was then placed in a water bath at 37°C and agitated for 5 minutes. The enzyme/calvaria mixture was decanted into a new conical tube and centrifuged to isolate the suspended cells. Fresh enzyme mixture was added between each digestion. The process was repeated twice, agitating for 15 minutes and then 25 minutes.

Following three complete digestions and subsequent resuspension in enzyme-free media, the cells from the second and third digestion were pooled in a conical tube. The resulting bone chips were added to the pooled suspension and hand-agitated for five seconds. The media was next decanted into a new conical tube and centrifuged to pellet the cells. The resulting isolated cells were then washed with Hanks Balanced Salt Solution four times (Cells were resuspended in the salt solution, pelleted and resuspended in Hanks Balanced Salt Solution for subsequent washing). Following 4 washings, the cells were resuspended in MEM with ES supplemented with 10% FBS and 10% Pen-Strep. The cells were then centrifuged at 4000 rpm and 25°C for 10 minutes.

The resuspended cells were counted using a hemacytometer, repeating four times to ensure accuracy.

Plating/Feeding

Cells were plated at 150,000 per flask and allowed to multiplied for four days in MEM supplemented with ES and 10% FBS with 1% Pen-Strep. Cells were fed 20 mL of this solution on the initial day following plating and 10 mL of this solution on subsequent days. In parallel, cells were also plated at 60,000 cells per 100 mm culture dish which were allowed to multiply for three days, fed by adding 3 mL of MEM with ES and 10% FBS with 1% Pen-Strep the first day, 10 mL of media the following day and 3 mL the remaining day.

Freshly isolated cells were fixed in 3.7% paraformaldehyde, rinsed with PBS and permeabilized prior to staining for cell shape. Actin fibers within the OBs were stained with Rhodamine-Phalloidin methanolic stock, while cell nuclei were highlighted using Hoechst stain. Under fluorescence with images overlayed, the actin fibers appear red and the nuclei blue.

pOBs were trypsinized after four days, counted and divided into groups of 1 million cells, then centrifuged at 4000 rpm and 25°C for 10 minutes in order to pellet the cells for subsequent ζ-potential analysis.

Alkaline phosphatase staining

To verify the identity of the cells, HA discs seeded with freshly isolated cells were stained for alkaline phosphatase (AP) activity 14 days after plating following the procedures established by Shu et al.[5] The cells were cultured using alpha-MEM supplemented with ES and 10% FBS, with 1% Pen-Strep added to prevent infection. Once the cells became confluent, a switch was made to using differential media (α-MEM with 10%FBS, 0.25% ascorbic acid and 0.25% β-glycerol phosphate).

Prior to staining for AP, the media was aspirated off and the culture wells rinsed with 1X PBS (pH 7.4). Cells were then fixed for 10 min using 2% paraformaldehyde. The fixed cells were next rinsed with and stored in 0.1 M cacodylic buffer (pH 7.4) at 4°C until ready to stain.

Cells were stained using a mixture of 11.8 mL distilled water 12.5 mL tris maleate buffer, 0.7 mL N,N dimethylformamide, 12.5 mg Naphthol AS-MX and 25 mg Fast Red. This mixture was filtered and enough added to cover each substrate. Cells were then incubated at 37°C for 30 minutes, or until a red color was visible. The stain buffer was then removed and the cells were rinsed once with cacodylic buffer.

ζ-potential measurement

pOB pellets were suspended in physiologic saline (0.154 M NaCl) at three different pH levels (7.3, 7.4, 7.4) coinciding with *in vivo* conditions, using complementary acid or base (1M HCl or 1MNaOH). pOB suspensions were transferred via 60 ml sterile syringes to the ESA chamber of an AcoustoSizer II (AZR II, Colloidal Dynamics, Warwick, RI) for ζ-potential analysis. Due to the greatly reduced specimen suspension volume (because of the difficulty in collecting large volumes of primary OBs), ζ-potential data were collected from a static suspension, yielding a single value for each of the three cell suspensions.

In addition, the ζ-potential of HA (sintered and unsintered) and β-TCP (unsintered) were also collected using ESA under N_2 purge and constant agitation, over a physiologically relevant pH range controlled via automated titration.

Finally, ζ-potential data were obtained for MC3T3-E1 clonal OBs, using static ESA measurement.[12]

RESULTS

Staining to show structure

Results of actin/Hoescht staining for primary rat OBs on dense hydroxyapatite vs. MC3T3-E1 OBs are shown in Figure 2. There are clear similarities between these two cell types, in that both cell types have an actin cytoskeleton (fibrous spiculated tendrils, stained red), forming an elongated cell body, bulging around the nucleus (circular nodes, stained blue) of each cell.

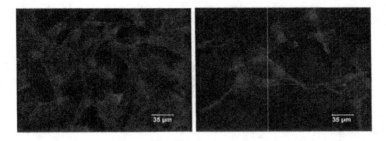

Figure 2 Primary fetal rat (left) osteoblast cells on dense hydroxyapatite and MC3T3-E1 clonal osteoblast cells (right) stained with rhodamine (actin fibers - red) and Hoescht (nuclei - blue) (400X).

AP assessment

The red color of the AP stain indicates the presence of alkaline phosphatase (AP). Alkaline phosphatase is an enzyme produced in the early stage of differentiation of osteoblast cells. The verification of AP demonstrates that the cells have begun to deposit bone mineral and have begun the early stages of differentiation.

AP results for the primary osteoblasts on dense HA discs showed AP activity on the surface of the discs. The red color was not consistently vivid over the entire area of the discs. This may have been because the cells lost viability during the digestion process and were not given a chance to recover before plating for a study. This study was plated directly after the digestion without plating on cell culture surfaces for growth. The inconsistent red color also may result if multiple cell types were collected from the calvaria. The cells used in previous studies were from a cloned cell line of osteoblasts, chosen for their proven ability to form calcified tissue *in vitro*.[6]

Previous studies have shown that MC3T3-E1 OBs stain for AP in similar fashion. A comparison is shown in Figure 3.

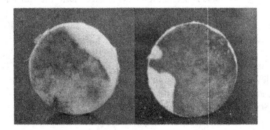

Figure 3 AP stain of primary rat osteoblast cells (left) and MC3T3-E1 clonal osteoblast cells (right) on dense HA discs.

ζ-potential comparison

ζ-potential values were determined as a function of pH (Figure 4) for OBs (MC3T3-E1, pOB) HA (sintered and unsintered) and β-TCP (unsintered). Over the range of pH 6–8, sintered HA exhibited ζ-potentials between 3.2–4.4 mV, unsintered HA had values between 12.1–17.8 mV and β-TCP showed values of 2-2.9 mV. From pH 7.3–7.5, MC3T3-E1 OBs showed ζ-

potentials between –29.4 and –52.4 mV and primary rat OBs exhibited values between –8.5 and –26.8 mV.

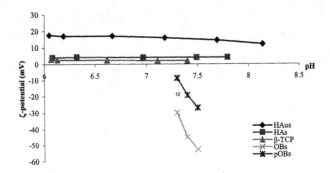

Figure 4 ζ-potential as a function of pH for MC3T3-E1 OBs (OBs), primary rat OBs (pOBs), HA (sintered and unsintered) and β-TCP (unsintered).

DISCUSSION/CONCLUSIONS

Our studies show the suitability of continued use of MC3T3-E1 OBs for use with bioceramic scaffolds such as hydroxyapatite for bone tissue engineering in terms of the clonal cell line's similar ζ-potential behavior. Preliminary AP stains performed in this study clearly indicate similarities in AP activity for both primary OBs and the MC3T3-E1 clonal OBs. However, further AP studies must be conducted to quantify the amount of AP activity present for OBs versus pOBs. AP is expressed during the early stages of differentiation, *in vivo* under normal bone healing and in the presence of HA, in which OBs produce an extracellular matrix. This phase is followed by increased osteocalcin expression and the formation of nodules and mineralization leading to the development of a bone-like structure.[7]

Factors such as differing species of origin, size and cell passage (the passage is one for primary OBs in comparison to passage 26 for MC3T3-E1 OBs), did not cause appreciable differences in the ζ-potential behavior between the pOBs and the MC3T3-E1 OBs, as shown in Figure 4. Both the pOBs and the MC3T3-E1 OBs have negative ζ-potentials over the physiologically relevant pH range examined in this study. The primary rat OBs in fact exhibit more than two times the negative ζ-potential over the same pH range in comparison to the MC3T3-E1 cells. So, while both types of OBs have a negative ζ-potential, making the use of the MC3T3-E1 cells a reliable indicator of the probability of OB attachment to positively charged scaffold materials, the enhanced negativity of the primary rat OBs may be related to differences in species (rat versus mouse) or from the presence of other cell types inadvertently included during the harvesting of the primary rat OBs.

Furthermore, the elimination of the problems inherent to primary cell cultures, such as the lengthy and meticulous dissection and digestion procedures, the difficulty in separating specific cell types and the requirements for new incubation environments, also support the use of clonal cell lines such as the MC3T3-E1 OB cell line as a convenient and effective alternative to primary cells for use in examining surface potential differences between cells and ceramic scaffolds used in bone tissue engineering.

REFERENCES
1. Mcintyre JP, Shackelford JF, Chapman MW, Pool RR (1991) Characterization of a bioceramic composite for repair of large bone defects. Ceram Bull 70:1499-1503
2. Hench LL (1991) Bioceramics: From concept to clinic. J Am Ceram Soc 74:1487-1510
3. Klein CP, Driessen AA, deGroot K, Hooff Avd (1983) Biodegradation behavior of various calcium phosphate materials in bone tissue. J Biomed Mater Res 17:769-784
4. Jarcho M (1981) Calcium phosphate ceramics as hard tissue prosthetics. Clin Ortho and Rel Res 157:259-278
5. Shu R, McMullen R, Baumann MJ, McCabe LR (2003) Hydroxyapatite accelerates differentiation and suppresses growth of MC3T3-E1 osteoblasts. J Biomed Mater Res 67A:1196-1204
6. Sudo H, Kodama HA, Amagai Y, Yamamoto S, Kasai S (1983) In vitro differentiation and calcification in a new clonal osteogenic cell line derived from newborn mouse calvaria. J Cell Biol 96:191-198
7. Quarles LD, Yohay DD, Lever L, Canton R, Wenstrup R (1992) Distinct proliferative and differentiated stages of murine MC3T3-E1 cells in culture: an in vitro model of osteoblast development. J Biomed Mater Res 7:683-692
8. O'Brien RW (1986) Electro-acoustic effects in a dilute suspension of spherical particles. J. Fluid Mech. 190:71-86
9. O'Brien RW (1990) The electro-acoustic equation for a colloidal suspension. J. Fluid Mech. 212:81-93
10. Oja T (1994) Matec Applied Sciences Inc. In, New York
11. Oppermann DA, (Baumann) MJC, Bement D (1998) In vitro stability predictions for the bone/hydroxyapatite composite system. J Biomed Mater Res 42:412-416
12. Smith IO, Soto MK, Baumann MJ, McCabe LR (2003) In vitro stability predictions of osteoblast interaction with hydroxyapatite and β-tricalcium phosphate. In: Ceramic transactions, Vol. 147; Bioceramics: Materials and applications IV. p 123
13. Yagiela JA, Woodbury DM (1977) Enzymatic isolation of osteoblasts from fetal rat calvaria. Anat. Rec. 188:287-306

EFFECT OF DOPANTS ON PROPERTIES OF NANOCRYSTALLINE HYDROXYAPATITE

Ashis Banerjee and Susmita Bose
School of Mechanical and Materials Engineering
Washington State University
Pullman, WA 99164-2920
E-mail: sbose@wsu.edu

ABSTRACT

Calcium phosphate based ceramics such as hydroxyapatite (HA) are widely used for bone implant applications due to their bioactivity and compositional similarity with natural bone. Low compressive strength and low fracture toughness of these compositions also prevent their application in load bearing components. Lowering the particle size of calcium phosphate based powders can improve densification and mechanical properties. This work presents synthesis and characterization of pure hydroxyapatite nanopowders using reverse micelle as a template and effect of dopant on the sintering behavior and mechanical properties. Calcium nitrate and phosphoric acid are used as sources of calcium-ion (Ca^{2+}) and phosphate-ion (PO_4^{3-}). Cyclohexane is used as oil phase and poly(oxyethylene)$_5$ nonylphenol ether (NP-5) and poly(oxyethylene)$_{12}$ nonylphenol ether are used as surfactants. Phase purity and specific average surface area analysis were done using x-ray diffraction method and BET surface area analysis. Effect of different synthesis parameters on morphology is studied by transmission electron microscopy (TEM). Sintering behavior and mechanical properties are studied using compacts of pure and alumina doped hydroxyapatite powders.

INTRODUCTION

Hydroxyapatite (HA, $Ca_{10}(PO_4)_6(OH)_2$), is an important biomaterial and is the principal inorganic constituent of bones and teeth [1]. Synthetic HA has excellent biocompatibility and bioactivity, and is widely used in many biomedical applications such as implants and coating onto prostheses [2-3]. For successful application of HA based ceramics in load bearing bone grafts, higher strength and toughness are necessary. Unfortunately, because of poor sinterability, HA ceramics show low strength and poor fracture toughness, especially in wet environment, which makes them unsuitable for even low load bearing applications. It is believed that nanostructured ceramics can improve the sinterability due to high surface energy and, therefore, improve mechanical properties [4]. But sintering behavior depends not only on the particle size but also depends on particle size distribution and morphology of the powder particles. Large particle size and wide particle size distribution along with hard agglomerates exhibits lower densification in HA [5]. Differential shrinkage between the agglomerates is also responsible to produce small cracks in the sintered HA [6]. Therefore synthesis of agglomerate free or soft agglomerated nanostructured HA is important to achieve high densification and consequently better mechanical properties.

Various synthesis methods of HA have been reported, including solid-state reaction [7], sol-gel synthesis [8], pyrolysis of aerosols [9], hydrothermal reaction [10] and microemulsion [4, 11-13]. The degree of success of these methods in preparing HA differs significantly from one to the other. Microemulsion has been shown to be one of the few techniques, which can give nanoparticles of HA with minimum agglomeration. A microemulsion is a thermodynamically stable transparent suspension of two immiscible liquids such as water and oil. The suspension is stabilized adding surfactant into the system. The microemulsion can be divided in two categories such as oil in water (micelle) and water in oil (reverse micelle). The micelle system is used in synthesis of mesoporous aggregates, whereas, reverse micelle template system is used in synthesis of nanoparticles with minimum agglomeration. In the case of reverse micelle system, the aqueous phase is dispersed as very small droplets surrounded by a monolayer of surfactant molecules in the continuous hydrocarbon phase. These small water droplets are stabilized in nonaqueous phase by surfactants act as a microreactor or nanoreactor in which reactions are conducted.

The present article describes synthesis of HA nanopowders using reverse micelle template system, and effect of synthesis parameters on powder particle morphology and distribution. Aqueous solution of calcium nitrate ($Ca(NO_3)_2$. $4H_2O$) and orthophosphoric acid (H_3PO_4) was used as water phase. The organic phase was prepared by mixing required amount of surfactant in cyclohexane. Two types of surfactant have been used in these experiments such as poly(oxyetheylene)$_5$ nonylphenol ether (NP5) and poly(oxyetheylene)$_{12}$ nonylphenol ether (NP12). Effect of aqueous to organic ratio, surfactant concentration, pH and effect of Al_2O_3 addition on the powder characteristics was studied. Powders were characterized by x-ray diffraction (XRD), BET specific surface area analysis and transmission electron microscopy (TEM). Sintering behavior and microhardness of the compacts were also characterized.

EXPERIMENTAL

Materials: Calcium nitrate ($Ca(NO_3)_2$. $4H_2O$, J. T. Baker, NJ), orthophosphoric acid (H_3PO_4, Fisher Scientific, NJ) were used as source of Ca^{2+} ion and PO_4^{3-} ion and alminum nitrate ($Al(NO_3)_3$. $9H_2O$, Alfa Aesar, MA) was used as source for Al_2O_3 dopant. Cyclohexane (Fisher Scientific, NJ) was used as organic solvent and poly(oxyehtylene)$_5$ nonylphenol ether (NP 5) and poly(oxyehtylene)$_{12}$ nonylphenol ether (NP 12) (Aldrich, WI) were used as surfactants. Ammonium hydroxide (J. T. baker, NJ) was used to adjust the pH of the emulsion system.

Synthesis of nano-powders: 0.5 M aqueous solution of Ca^{2+}-ion was prepared by dissolving 2.361 g amount of $Ca(NO_3)_2$. $4H_2O$ in distilled water. 0.686 g of phosphoric acid was added to the system to maintain Ca to P ratio of 1.67:1. The organic phase was prepared by adding required amount surfactant in the cyclohexane with vigorous stirring. To study the effect of aqueous to organic ratio on the powder characteristics, 10 vol% surfactant was dissolved in cyclohexane and the aqueous phase and the organic phase were mixed in the ratio of 1:10, 1:20 and 1:30 with vigorous stirring. The effect of pH was studied by adjusting the pH of the emulsion to 7 and 9 with dropwise addition of NH_4OH. All reactions were aged for 24 h at room temperature. After aging, the emulsion was evaporated on hot plate at 150°C followed by complete drying at 450°C. Dry precursor powder was calcined at 650°C for 2 h to get carbon free crystalline HA nano-powder. 2.5 wt% Al_2O_3 doped HA nano-powder was prepared by adding required amount of $Al(NO_3)_3 \cdot 9H_2O$ in the aqueous solution of $Ca(NO_3)_2 \cdot 4H_2O$ and H_3PO_4 during synthesis.

Characterization: Powders were characterized for their phase purity by x-ray diffraction (XRD) analysis using a Philips PW 3040/00 X'pert MPD system at room temperature with Co-K$_\alpha$ radiation and a Ni-filter over the 2θ ranges of $20°$ to $70°$ at a step size of $0.02°$ (2θ) and a count time of 0.5 sec per step. The specific average surface area measurements were done using a five point BET surface area analysis (Tristar 3000, Micromeritics). Powder morphology and particle size was evaluated using a transmission electron microscope (JEOL, JEM 120). Sintering behavior of these powders were studied using uniaxially pressed compacts. Compacts were sintered at 1250 and 1400°C. Microhardness measurements were done using a Vickers microhardness tester.

RESULTS AND DISCUSSION

Surfactants form reverse micelle aggregate in organic solvents beyond a certain concentration. These aggregates can be stabilized in presence of small amount of water. They can be formed both in presence and in absence of water. In organic solvents, polar head groups of surfactants form small cores, which can be stabilized in the presence of small amount of water. If the medium is water free, the aggregates are very small and polydisperse. Water is readily solubilized in the polar core and the size of the polar core depends on the amount of water present in the system. Figure 1 shows the formation of a reverse micelle system.

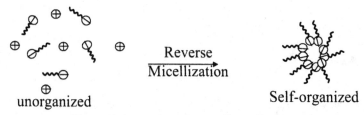

Figure 1. Formation of reverse micelle

When NP 5 or NP 12 was added in cyclohexane, it forms small polar cores by organizing the polar head groups away from non polar organic solvents. When aqueous solutions of $Ca(NO_3)_2$ and H_3PO_4 were added in that organic phase the water goes to the small polar cores and forms microreactors. HA is stable only above pH 4.2. All reactions were carried out at pH 7 and aged for 24 h at room temperature. The aging time had a significant role on chemical reaction and the degree of crystallinity. It was observed that lowering the aging time might increase BET specific average surface area of the powder but decreased the degree of crystallintiy. The HA powder with lower crystallinity may decompose to tri calcium phosphate (TCP) at higher temperature during sintering. Therefore we have conducted our research at 24 h aging time to allow sufficient time for the reaction and increase the degree of crystallinity.

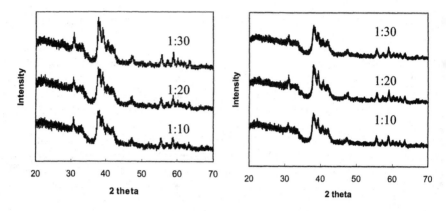

Figure 2. XRD plot of synthesized HA at different aqueous to organic ratio (1:10, 1:20 and 1:30) a) by using NP 5 at pH 7 and b) by using NP 5 at pH 9.

Figure 2 shows the X-ray diffraction pattern of HA nanopowders synthesized by using NP 5 surfactant. Figure 2a shows the effect of aqueous to organic ratio on the phase formation of HA at pH 7 and Figure 2b shows the same at pH 9. It was observed that aqueous to organic ratio does not have any significant effect on the HA phase formation. But the degree of crystallinity changes by changing pH of the synthesis medium from 7 to 9.

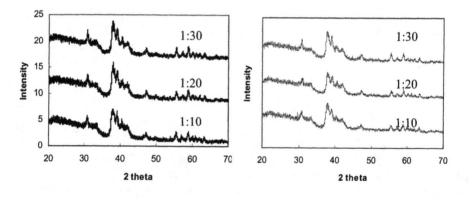

Figure 3. XRD plot of synthesized HA at different aqueous to organic ratio (1:10, 1:20 and 1:30) a) by using NP 12 at pH 7 and b) by using NP 12 at pH 9.

Figure 3 shows x-ray diffraction plot of pure HA using NP 12 surfactant. Figure 3 a shows the effect of aqueous to organic ratio on the phase formation of HA at pH 7 and Figure 3b shows the same at pH 9. In both cases, phase pure HA nanopowders were obtained. Therefore the aqueous to organic ratio does not have any significant effect on phase formation.

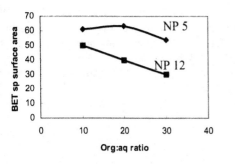 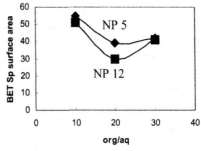

Figure 4. Effect of aqueous to organic ratio on specific surface area a) at pH 7 and b) at pH 9.

Figure 4 shows the effect of aqueous to organic ratio and pH on the BET specific average surface area. Figure 4a shows the effect on the BET surface area using NP 5 and NP 12 at pH 7 and Figure 4b shows the effect on BET surface area by using NP 5 and NP 12 at pH 9. It was observed that at pH 7, increasing organic to aqueous ratio decreased the BET surface area. But NP 5 always formed higher surface area powders compared to NP 12. The BET surface area analysis at pH 9 shows increasing organic to aqueous ratio first decreased the surface area and then increased. Further research is going on to understand the mechanism of these types of behavior.

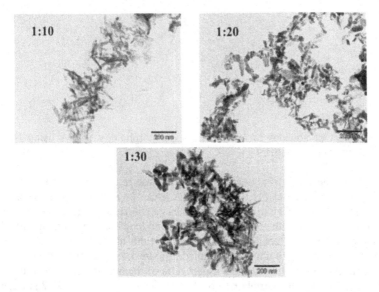

Figure 5. Effect of aqueous to organic ratio on morphology of HA using NP 5 and powder was synthesized at pH 7.

Figure 5 shows the effect of aqueous to organic ratio on the morphology of the HA nanoparticles made at pH 7 using NP 5 surfactant. It can be seen that increasing organic phase decrease the aspect ratio of the powders. At lower organic content, high aspect ratio powders were formed with average width of 10-20 nm and length about 100-120 nm. As the organic content in the system increased the aspect ratio of the powder particles decreased and formed platelet type particles.

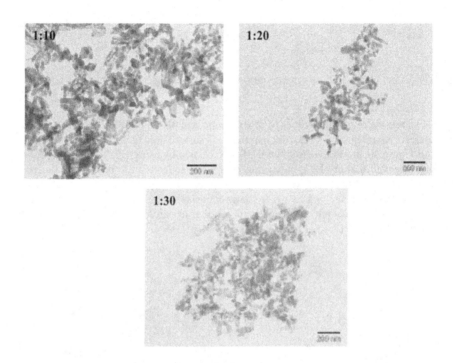

Figure 6. Effect of aqueous to organic ratio on morphology of HA using NP 5 and powder was synthesized at pH 9.

Figure 6 shows the effect of aqueous to organic phase ratio on the morphology of HA nanoparticles synthesized at pH 9 using NP5. It can be seen that increase in organic phase decreases the aspect ratio of the powder. At lower organic content (1:10), the width of the particles is in the range of 20-30 nm and length in the range of 100-125 nm. But at higher organic content (1:30), the particle width in the range of 20-30 nm and length in the range of 50-70 nm were obtained. Figure 5 and figure 6 indicates that with the same aqueous to organic ratio, the aspect ratio of the powder is lower in the case of pH 9 than pH 7. Also the particle size of the powder synthesized at pH 7 is larger than the powder synthesized at pH 9.

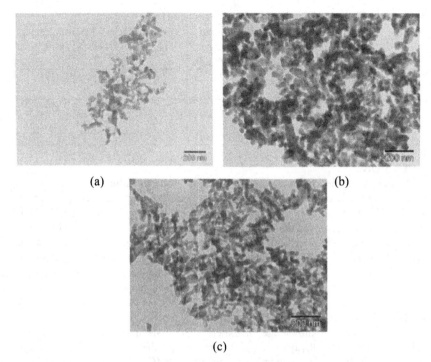

(a)   (b)

(c)

Figure 7. TEM analysis of HA powder synthesized by using NP 5 and at pH 9 a) 10 vol% NP 5, b) 20 vol% NP 5 and c) 2.5 wt% Al_2O_3 doped HAP by using 20 vol% NP 5

Figure 7 shows the effect of increase in surfactant concentration on powder morphology. It was observed that as surfactant concentration increases aspect ratio of HA nanopowder decreases. Figure 7c shows that addition of alumina increases the aspect ratio of the powder compared to pure HAP.

Addition of alumina does not have any detrimental effect on HA phase formation. Phase pure HA powder was obtained after 2.5wt% alumina addition. The BET specific average surface area also increased with the addition of alumina. Uniaxially pressed compacts made of pure and alumina doped HA powder was sintered at 1250 and 1400°C. Density and microhardness of these compacts are shown in Table 1. At 1250°C higher densification (97%) was achieved in case of compacts made of pure HA nanopowder compared to that of Al_2O_3 doped HA. Aspect ratio of the pure HA was lower compared to Al_2O_3 doped HA. Low aspect ratio powders have high packing efficiency and were responsible for high densification of pure HA at 1250°C. At 1400 °C, Al_2O_3 doped powders sintered better, but pure HA degrades because of the formation of other calcium phosphate phases.

Table 1. Properties of pure and 2.5 wt% Al$_2$O$_3$ doped HA

Sample	BET (m^2/g)	Sintered density at 1250°C/2h	Sintered density at 1400°C/2h	Microhardness (1250°C)	Microhardness (1400°C)
HA	35	97	99	6 GPa	4.4 GPa
Al$_2$O$_3$ doped HA	50	80	98	-	6.2 GPa

CONCLUSIONS

Nanocrystalline hydroxyapatite powders were synthesized using emulsion method. Nature of surfactant, aqueous to organic ratio and pH of the system had significant effect on the specific surface area and morphology of the particles. Increase in organic content decreased the aspect ratio of the powder. Addition of Al$_2$O$_3$ increases BET specific average surface area of the HA powder. TEM study showed that addition of Al$_2$O$_3$ increased aspect ratio of particles. Pure HA nanoparticles showed 97% densification at 1250°C via pressureless sintering. Under same condition, the compacts made of 2.5 wt% Al$_2$O$_3$ doped HA powders showed 80 % densification.

ACKNOWLEDGEMENTS

Authors would like to acknowledge the experimental assistance and valuable suggestions from Prof. A. Bandyopadhyay. Financial support from the National Science Foundation (NSF) under the PECASE award to Prof. Susmita Bose (Grant No. CTS-0134476) is also acknowledged.

REFERENCES

1) L. L. Hench, "Bioceramics: from concept of clinic," *Journal of the American Ceramic Society*, **74** 1487-1510 (1991).

2) K de Groot, "Bioceramics consisting of calcium phosphate salts," *Biomaterials*, **1** 47-50 (1980).

3) F. H. Lin, C. C. Lin, H. C. Lu, Y. Y. Huang, C. Y. Wang and C. M. Lu, "Sintered Porous DP-Bioactive Glass and Hydroxyapatite as Bone substitute," *Biomaterials*, **15** 1087-1098 (1994).

4) S. Bose and S. K. Saha, "Synthesis and Characterization of Hydroxyapatite Nanopowders by Emulsion Technique," *Chemistry of Materials*, **15** 4464-4469 (2003).

5) M. G. S. Murray, J. Wang, C. B. Ponton and P. M. Marquis, "An Improvement in Processing Hydroxyapatite Ceramics," *Journal of Materials Science*, **30** 3061-3074 (1995).

6) A. J. Ruys, M. Wei, C. C. Sorrel, M. R. Dickson, A. Brandwood and B. K. Milthpore, "Sintering effects on the Strength of Hydroxyapatite," *Biomaterials*, **16** 409-415 (1995).

7) R. R. Rao, H. N. Roopa and T. S. Kanan, "Solid State Synthesis and Thermal Stability of HAP and HAP - β TCP Composite Ceramic powders," *Journal of Materials Science: Materials in Medicine*, **8** 511-518 (1997).

8) A. Deptula, W. lada, T. Olczak, A. Borello, C. Alvani and A. Dibartolomeo, "Preparation of Spherical Powders of Hydroxyapatite by Sol-Gel Process," *Journal of Non Crystalline Solids*, **147** 537-541 (1992).

9) M. Valet-Regi, M. T. Gutierrez-Rios, M. P. Alonso, M. I. De Frutos and S. Nicolopoulos, "Hydroxyapatite Particles Synthesized by Pyrolysis of Aerosols," *Journal of Solid State Chemistry*, **112** 58-64 (1994).

10) H. Hattori and Y. Iwadate, "Hydrothermal Preparation of Calcium Hydroxyapatite Powders," *Journal of the American ceramic Society*, **73** 1803-1805 (1990).

11) G. K. Lim, J. Wang, S.C. Ng, C. H. Chew and L. M. Gan, "Processing of Hydroxyaptite via Microemulsion and Emulsion Routes," *Biomaterials*, **18** 1433-1439 (1997).

12) S. Bose and S. K. Saha, "Synthesis of hydroxyapatite nanopowder using sucrose templated sol-gel method" *Journal of the American Ceramic Society*, **86** [6], pp. 1055-57 (2003).

13) K. Sonoda, T. Furuzono, D. Walsh, K. Sato and J. Tanaka, "Influence of Emulsion on Crystal Growth of Hydroxyapatite," *Solid State Ionics*, **151** 321-327 (2002).

14) J. Liu, X. Ye, H. Wang, M. Zhu, B. Wang and H. Yan, "The Influence of pH and Temperature on the Morphology of Hydroxyapatite Synthesized by Hydrothermal Method," *Ceramics International*, **29** 629-633 (2003).

LOW-TEMPERATURE SYNTHESIS OF HA-SEEDED TTCP (Ca₄(PO₄)₂O) POWDERS AND THEIR *IN VITRO* APATITE-INDUCING ABILITY AT 37°C AND pH 7.4

LOW-TEMPERATURE SYNTHESIS OF HA-SEEDED TTCP ($Ca_4(PO_4)_2O$) POWDERS AND THEIR *IN VITRO* APATITE-INDUCING ABILITY AT 37°C AND pH 7.4

Sahil Jalota, A. Cuneyt Tas, and Sarit B. Bhaduri
School of Materials Science and Engineering, Clemson University, Clemson, SC 29634, USA

ABSTRACT

This paper reports the synthesis of hydroxyapatite (HA: $Ca_{10}(PO_4)_6(OH)_2$)-seeded (5 to 6 wt%) TTCP ($Ca_4(PO_4)_2O$) powders at 1230°C. This temperature is apparently much lower than the typical temperature of synthesis (i.e., 1350° to 1500°C) reported so far for TTCP powders. The starting materials used in our powder synthesis route were Ca-acetate monohydrate ($Ca(CH_3COO)_2 \cdot H_2O$) and ammonium dihydrogen phosphate ($NH_4H_2PO_4$). HA-seeded TTCP powders were tested for their apatite-inducing ability by soaking them in synthetic body fluid (SBF) solutions at 37°C from 36 to 96 hours. Bioactive TTCP powders of this study readily covered with carbonated apatitic calcium phosphates within the first 72 hours. Crystal structure similarity between TTCP and HA, more specifically, the presence of apatitic layers within the unit cell of TTCP causes this *in vitro* bioactivity.

INTRODUCTION

Brown and Epstein [1] previously showed that tetracalcium diphosphate monoxide, $Ca_4(PO_4)_2O$ (TTCP), has a structural relationship to HA; $Ca_{10}(PO_4)_6(OH)_2$, which may be considered to be the idealized form of the major inorganic phase of human bone. The more commonly used name for $Ca_4(PO_4)_2O$ is tetracalcium phosphate (from the formula $4CaO \cdot P_2O_5$). TTCP crystallizes in the monoclinic unit cell a = 9.462, b = 11.965, c = 7.023 Å, β= 90.9° in space group $P2_1$ with $4[Ca_4(PO_4)_2O]$ per cell. It has a density of 3.06 g/cm³ [2]. TTCP melts at around 1710°C [3]. The magnitudes of a and c parameters of the unit cells of HA and TTCP are comparable to one another. It is also interesting to note that when one multiplies the unit cell volume of HA by a factor of 1.5, one obtains the exact value of the unit cell volume of TTCP. $Ca_4(PO_4)_2O$ is also known to display extensive twinning, and suspected to have a high-temperature orthorhombic polymorph [4]. The positions of the P atoms and the Ca and oxide ions lie close to those required by an orthorhombic space group, *Pmcn*. This is a likely explanation for the appreciable twinning exhibited by $Ca_4(PO_4)_2O$ [5].

How can TTCP transform into apatite? $Ca_4(PO_4)_2O$ contains a crystallographic layer similar to an "apatitic layer" present in both $Ca_{10}(PO_4)_6(OH)_2$ and $Ca_8H_2(PO_4)_6 \cdot 5H_2O$ (octacalcium phosphate, OCP). Because of this layer, an epitaxic relationship between TTCP and HA, as well between TTCP and OCP, may be present [5]. Epitaxy, the ordered growth of one substance on another, is often thought of as being governed chiefly by the metric fits of crystallographic networks based on the unit cell translations [5]. The combination of good metric and chemical fits between HA and OCP may cause the epitaxy between these two compounds to occur [6]. Undetected epitaxy between HA and TTCP would increase the Ca/P molar ratio of an "apparent" apatite above 1.667, and on the microscopic scale the material would resemble solid solution. HA is, therefore, an unusual substance in that it has two related salts with which it may enter into epitactic relationships. One salt, $Ca_8H_2(PO_4)_6 \cdot 5H_2O$, is more acidic, and the other, $Ca_4(PO_4)_2O$, is more basic [5].

Production of $Ca_4(PO_4)_2O$ in an aqueous environment is extremely difficult, if not impossible, because of the oxide ion. TTCP is the most soluble compound among all the calcium phosphates. In aqueous solutions hydroxyl ions incorporate themselves into the formed precipitates. A carbonate- or hydroxyl ion-containing product will form more easily. For this reason, synthesis of TTCP, which is nowadays used as one of the main components of self-setting orthopedic and dental cements, has only been limited to solid state reactive firing (SSRF) [7-9]. Tetracalcium phosphate was typically synthesized by the solid-state reaction of calcium carbonate ($CaCO_3$) and dicalcium phosphate anhydrous ($CaHPO_4$) powders, ball-milled with one another at a Ca/P molar ratio of 2.0. Powders prepared by Chow et al. [7-9] were usually heated at 1450° to 1500°C for 6 to 12 h, and then quenched to room temperature. Slow cooling of TTCP, instead of quenching, from high temperatures results in the formation of undesired secondary phases, such as HA, CaO, $CaCO_3$, and β-$Ca_3(PO_4)_2$. As of now, it was impossible to obtain TTCP without quenching from a temperature in excess of 1300°C. Rapid cooling is essential in order to prevent the formation of HA and calcium oxide, a side reaction that takes place between 800° and 1250°C in moist atmospheres [10]. Quenched, sintered body of tetracalcium phosphate was needed to be grind to a particle size less than 45 μm, especially if such powders were to be used in cement applications [9].

Sargin et al. [11] have studied the influence of different starting materials on the synthesis of TTCP powders. In this important study, Sargin et al. [11] have shown that if one starts with $CaCO_3$ and $NH_4H_2PO_4$ powder mixtures, it is possible to decrease the temperature of synthesis of single-phase TTCP down to about 1350°C. They also carefully analyzed the experimental X-ray and infrared data (IR) of their samples and tabulated the band assignments for TTCP for the first time. What Sargin et al. basically accomplished [11] was to separate Ca and P from one another in their reactants. In Chow's work [7-9, 12], in stark contrast with Sargin et al., the calcium needed to form TTCP was taken from both $CaCO_3$ and $CaHPO_4$. During heating, $CaCO_3$ decomposes by evolving CO_2 at around 900°C, and $CaHPO_4$ first transforms into $Ca_2P_2O_7$ [13]. Therefore, in Chow's approach, high temperatures were indeed necessary to convert the pyrophosphate ions into orthophosphates before starting to form the TTCP phase.

The discovery that the basic TTCP reacts with acid dicalcium phosphate anhydrous $CaHPO_4$ to form pure hydroxyapatite led to the development of a novel self-setting calcium phosphate cement (CPC) by Brown and Chow in 1985 [7, 8, 14]. TTCP-containing self-setting calcium phosphate cements are now commercially available [15, 16] and have a proven track record in clinical/surgical use.

The present study investigates a new chemical synthesis route to prepare 5 to 6 wt% HA-seeded TTCP powders. One major aim is to decrease the synthesis temperature below that reported by Sargin et al. [11], i.e., 1350°C. The powder synthesis route presented here consisted of physical mixing of stoichiometric amounts of calcium acetate monohydrate ($Ca(CH_3COO)_2 \cdot H_2O$) and ammonium dihydrogen phosphate ($NH_4H_2PO_4$) powders, followed by heating at 1230°C in air for 12 h and quenching. This process yielded finely divided powders at the lowest temperature yet reported for TTCP synthesis. Biomimetic precipitation of carbonated hydroxyapatite at 37°C and pH 7.4 on these powders is tested by immersing these into a synthetic body fluid (SBF) from 36 to 96 hours [17].

EXPERIMENTAL PROCEDURE

Powder synthesis: HA-seeded TTCP powders were synthesized in small batches. First, 0.77 g of $NH_4H_2PO_4$ was ground into a very fine powder by using an agate mortar and pestle. 2.45 g of $Ca(CH_3COO)_2 \cdot H_2O$ was then added to the powder. Two powders were dry mixed in the mortar for about 45 minutes, the powder mixture was heated at 300°C in air for 30 minutes, followed by re-mixing in a mortar for 15 minutes. Powders were then heated at 800°C in air for 1 hour, followed by homogenization by further grinding. These powders had a greyish/bluish tint and were heated (with a heating rate of 2°C/min) at 1230°C in air for 6 hours using a $MoSi_2$ box furnace. Following firing the red-hot alumina boat containing the powders was taken out, and placed into a desiccator. Obtained powders were lightly ground and homogenized in an agate mortar for 10 minutes, and re-fired at 1230°C for another 6 hours, followed by quenching to RT. Powders obtained were fluffy and easily ground into a fine powder after few minutes of manual grinding in an agate mortar. This procedure is suitable for the synthesis of HA-seeded TTCP powders. Caution was taken not to overgrind the obtained TTCP powders, since they are prone to transform into HA even by a small presence of mechanochemical activation.

In vitro testing: Apatite-inducing ability of HA-seeded TTCP powders was tested by immersion of the powders in a synthetic body fluid (SBF) solution. The details of preparing these solutions have been previously reported [17]. Briefly, the SBF solution used was a *tris*/HCl-buffered, 27.0 mM HCO_3^- ion-containing solution with the following ion concentrations: 2.5 mM Ca^{2+}, 1 mM HPO_4^{2-}, 27 mM HCO_3^-, 142 mM Na^+, 5 mM K^+, 1.5 mM Mg^{2+}, 0.5 mM SO_4^{2-}, 125 mM Cl^-, *tris*-buffered at pH=7.4. The solutions hereby used had the same carbonate ion concentration with that of human blood plasma. In vitro tests were performed in 90 mL-capacity glass bottles, which contained 2.0 grams of HA-seeded TTCP powders and about 90 mL of SBF solution. Tightly sealed bottles were placed into an oven at 37±1°C and kept undisturbed in that oven over the entire sampling times of 36, 72 and 96 hours. At the end of the stated immersion periods, partly solidified samples were washed with deionized water, and dried overnight at 37°C.

Characterization: Phase assemblage of powder samples was analyzed using a powder X-ray diffractometer (XDS-2000, Scintag Corp.), operated at 40 kV and 30 mA, equipped with a Cu-tube, using a step size of 0.01° 2θ. Fourier-transformed infrared reflective spectroscopy (Nicolet 550, Thermo Corp.) analyses were performed on the powder samples using a diamond ATR holder. Thermogravimetric analyses (TGA, Mettler Corp.) were performed in air on the starting chemicals of our powder synthesis over the range of 20 to 740°C, with a scan rate of 5°C/min. Surface morphology of the samples was studied using a Field-emission SEM (S-4700, Hitachi) after coating the samples for 4 to 6 minutes with Cr or Au.

RESULTS AND DISCUSSION

Thermogravimetric data for the starting chemicals of powder synthesis are shown in Figure 1. $NH_4H_2PO_4$ melts at around 200°C, and forms an acidic liquid. This acidic liquid immediately reacts with Ca-acetate, which until 200°C has already lost its water of hydration. The purpose of heating the powder mixture first to 300°C and then grinding the precursors at this stage is the homogenization of the reaction products of molten ammonium dihydrogen phosphate and calcium acetate. Upon continued heating to 800°C, volatilization of NH_3 from $NH_4H_2PO_4$, followed by its condensation to highly-reactive P_2O_5 is completed, whereas acetone, carbon dioxide and water vapor would be the main decomposition products of Ca-acetate. An amorphous intermediate (consisting of long chains of condensed phosphate of the form $(PO_3)_n$) will form as a result of these reactions, which will be converted into partially crystalline β-$Ca(PO_3)_2$ with an increase in temperature [18]. At this stage, presence of small amounts of $Ca_2P_2O_7$ may also be expected, however, this was not detected in this study. TTCP powders obtained after 2nd step of 1230°C firing/quenching has the characteristic XRD diagram given in Figure 2. These powders contained around 5 to 6 wt% of HA phase. Owing to the overall stoichiometry of the initial powder mixtures, detection of HA in the powders should automatically call for a small presence of $Ca(OH)_2$ or CaO in them. Major peaks of CaO and $Ca(OH)_2$ are around 37.36° and 34.17° 2θ, respectively, which are probably overlapping by reflections from the TTCP phase. Most HA peaks also overlap with those of TTCP, and the only peak of HA is encountered at 25.88° 2θ.

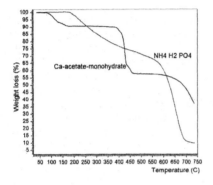
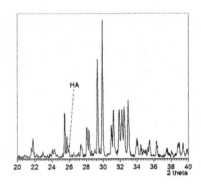

Fig. 1 TG traces of the starting chemicals   Fig. 2 XRD data of HA-seeded TTCP

The addition of HA seeds into TTCP powders have previously been tested by Hamanishi et al. [19], however that study intentionally added about 40 wt% precipitated HA into a cement mixture of $TTCP-CaHPO_4\cdot2H_2O$. The level of HA-seeding in the present study is much lower than this. Particle morphology of the HA-seeded TTCP powders are given in Figures 3 and 4 below.

Figs. 3 and 4 (*l-to-r*)   Morphology of the HA-seeded TTCP powders synthesized at 1230°C

Powders displayed a unique vermicular microstructure, which was indicative of the vapor- and liquid-phase reactions taking place. The remnants of the amorphous-looking chunks of condensed phosphates were observed in the form of isolated larger chunks. Good handling properties (between the fingers of an experienced self-setting cement practitioner) when wetted with a small amount of water (L/P = 0.1 to 0.2 mL/g), of these powders can be ascribed to this microstructure. A decrease in the average particle size of a cement powder typically leads the way to using lesser volumes of setting starter solution, which means reaching a higher compressive strength in the studied cement system [20]. On the other hand, TTCP powders synthesized at 1450° or 1500°C, using the starting chemicals $CaCO_3$ and $CaHPO_4$, do not possess such fine particle morphology, and they need grinding of the sintered ceramics to particle sizes less than 40-45 μm [9, 21]. It has recently been reported by Gbureck *et al.* [22] that ball milling a TTCP powder synthesized at 1500°C even in pure ethanol ensures its complete transformation into HA in 24 hours. This obviously means that ball milling of TTCP strongly degrades its cement properties. The same concern is also applicable to the most versatile constituent of self-setting calcium phosphate cement formulations, i.e., α-TCP [23].

FTIR pattern of the HA-seeded TTCP powders is depicted in Figure 5. This experimental pattern precisely matches those previously reported for tetracalcium phosphate by Sargin *et al.* [11] and Posset *et al.* [24], respectively. Vibrational wavenumbers were assigned by the above-mentioned researchers. In the data of Figure 5, threefold degenerate stretching (1105-989 cm^{-1}) mode (v_3), symmetric stretching (962-941 cm^{-1}) mode (v_1), and threefold degenerate deformational (620-571 cm^{-1}) mode (v_4) vibrations were all visible [11, 24]. The inset in Fig. 5 displays the characteristic OH stretching vibration (3650 cm^{-1}), originating from $Ca(OH)_2$ [11]. On the other hand, the weak band at 872 cm^{-1} belongs to carbonate groups, and those were considered to be associated with the HA present.

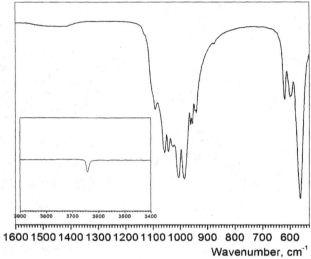

Fig. 5 FTIR trace of HA-seeded TTCP powders

SBF solutions are able [25] to induce apatitic calcium phosphate formation on metals, ceramics or polymers (with proper surface treatments). It is also known that an amorphous calcium phosphate (ACP) precursor is always present during the precipitation of apatitic calcium phosphates from highly supersaturated solutions [26]. Posner $et\ al.$ [27] proposed that the process of ACP formation in solution involved the formation of $Ca_9(PO_4)_6$ clusters which then aggregated randomly to produce the larger spherical particles or globules, with the intercluster space filled with water. Such clusters, we believe, are the transient solution precursors to the formation of carbonated globules with the nominal stoichiometry of $Ca_{10-x}(HPO_4)_x(PO_4)_{6-x}(OH)_{2-x}$, where x might be converging to 1 [28]. Onuma $et\ al.$ [29] have demonstrated, by using dynamic light scattering, the presence of calcium phosphate clusters from 0.7 to 1.0 nm in size even in clear simulated body fluids. They reported the presence of calcium phosphate clusters in SBF even when there was no visible precipitation. Therefore, one only needs some suitable surface to be immersed into an SBF solution, which would likely trigger the hexagonal packing [29] of those nanoclusters into globules or spherulites of apatitic calcium phosphates. SEM photomicrographs given in Figures 6 through 9 below exhibit the rapid formation of those apatitic calcium phosphate globules after soaking the HA-seeded TTCP powders in SBF for 36, 72, and 96 hours at 37°C, respectively. These are bioactive powders.

Fig. 6　**36 h** in SBF at 37°C　　　　　Fig. 7　**72 h** in SBF at 37°C

Fig. 8　**96 h** in SBF at 37°C　(*high mag*)　　　Fig. 9　96 h in SBF at 37°C　(*low mag*)

Figure 6 reveals the twinning in the TTCP grains, which was first mentioned by Dickens *et al.* [5] in a purely crystallographic study, more than three decades ago. In a supersaturated (with respect to apatite formation) solution, such as SBF, the apatitic layers of TTCP readily react with the calcium phosphate globules (Fig. 6), and within the subsequent 36 hours the external surfaces of the TTCP grains totally transform into apatite (Fig. 7). Why do we say "the external surfaces or skins?" Simply because the XRD and FTIR data of these samples (as given below in Figures 10 and 11) affirmed that the bulk of these samples was still TTCP. As a function of increasing soaking time in SBF, the overall XRD peak intensities were inclined to slightly decrease, with an accompanying significant increase in the intensities of the HA reflections (Fig. 10, * indicates HA peaks). FTIR data of Figure 11 (where * indicates carbonate groups) showed the appearance of a pronounced band at 1420 cm^{-1} in the high-energy C–O region along with a well-defined band at 872 cm^{-1}, known to be specific for a carbonated apatite [30]. OH bands were seen at 3571 cm^{-1} after soaking in SBF.

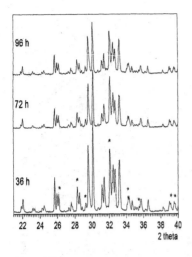

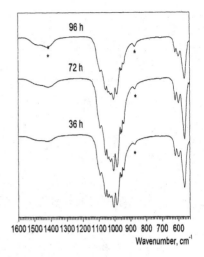

Fig. 10   XRD traces of SBF-soaked samples   Fig. 11   FTIR traces of SBF-soaked samples

We have also noted that the surface pH values of the as-synthesized HA-seeded TTCP powders were in the vicinity of 11.8 (measured after 15 minutes of adding a 50 mg portion of dry powder to 10 mL of deionized water to form a suspension, while a pH electrode is being inserted into this suspension). Such high pH values observed in such TTCP powders may cause cell *necrosis*, in case they were directly implanted *in vivo*. However, as shown in Figure 12 below, with an increase in soaking time in an SBF solution (at 37°C) of pH 7.4, pH starts decreasing. This is ascribed to the transformation of HA-seeded TTCP powders into carbonated apatitic calcium phosphates formed under biomimetic conditions provided *in vitro* by the SBF solutions. On the other hand, for comparison purposes, pure, stoichiometric, crystalline hydroxyapatite has a pH value of about 9.5. After 96 hours of soaking in SBF, recorded pH value was 9.56. It is reasonable to assume at this point that this pH value would have continued to drop toward the physiological pH, if we had continued the soaking beyond 4 days [31, 32]. The pH diagram of TTCP powders given in Figure 12 may very well serve the practitioner to precisely tailor the "amount of an acidic calcium phosphate (DCPA, DCPD, etc.) that needs to be blended" with these powders in designing a TTCP-based cement formulation having a neutral pH at the very start of the cement-forming process.

Being the calcium phosphate compound of highest solubility at the physiological pH, the hydrolysis of TTCP should be regarded as a process of initial dissolution followed by the precipitation of a less soluble phase. In an SBF environment (i.e., 37°C and pH 7.4), the precipitating phase can only be nanosize, carbonated hydroxyapatite, i.e., the bone mineral. The presence of HA seeds in TTCP simply accelerates this *in vitro* biomineralization process.

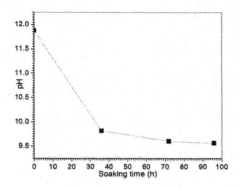

Fig. 12   Change in surface pH of HA-seeded TTCP in SBF

CONCLUSIONS

(1) 5 to 6 wt% HA-seeded TTCP ($Ca_4(PO_4)_2O$) powders were synthesized at 1230°C, by using the starting chemicals of calcium acetate monohydrate and ammonium dihydrogen phosphate, followed by quenching from the synthesis temperature.

(2) This is the lowest temperature ever reported for the manufacture of TTCP powders.

(3) HA-seeded TTCP powders can readily be used in the production of self-setting calcium phosphate cements, after physical blending of those with an acidic calcium phosphate salt, such as DCPA, DCPD, OCP or monocalcium phosphates.

(4) HA-seeded TTCP powders were shown to have a very strong apatite-inducing ability when they are brought into contact with SBF solutions of pH 7.4 at 37°C, within the first 36 to 96 hours.

ACKNOWLEDGMENTS

This work was partially supported by NSF DMII 0085100.

REFERENCES

[1]W. E. Brown and E. F. Epstein, *J. Res. Natl. Bur. Stand.*, **69A** 547-551 (1965).
[2]H. Bucking and G. Linck, *Stahl Eisen*, **7** 245-249 (1887).
[3]G. Tromel and W. Fix, *Arch. Eisenhuttenwes.*, **32** 219-212 (1961).
[4]G. Tromel and C. Zaminer, *Arch. Eisenhuttenwes.*, **30** 205-209 (1959).
[5]B. Dickens, W. E. Brown, G. J. Kruger, and J. M. Stewart, "$Ca_4(PO_4)_2O$, Tetracalcium Diphosphate Monoxide. Crystal Structure and Relationships to $Ca_5(PO_4)_3(OH)$ and $K_3Na(SO_4)_2$," *Acta Cryst.*, **B29** 2046-2056 (1973).
[6]X. Lu and Y. Leng, "TEM Study of Calcium Phosphate Precipitation on Bioactive Titanium Surfaces," *Biomaterials*, **25** 1779-1786 (2004).
[7]W. E. Brown and L. C. Chow, "Dental Restorative Cement Pastes," *US Patent No. 4,518,430*, May 21, 1985.
[8]L. C. Chow and S. Takagi, "Self-setting calcium phosphate cements and methods for preparing and using them," *US Patent No. 5,525,148*, June 11, 1996.
[9]Y. Matsuya, S. Matsuya, J. M. Antonucci, S. Takagi, L. C. Chow, and A. Akamine, "Effect of Powder Grinding on Hydroxyapatite Formation in a Polymeric Calcium Phosphate Cement Prepared from Tetracalcium Phosphate and Poly(methyl vinyl ether maleic-acid)," *Biomaterials*, **20** 691-697 (1999).

[10]H. Monma, M. Goto, H. Nakajima, and H. Hashimoto, "Preparation of Tetracalcium Phosphate," *Gypsum Lime*, **202** 151-155 (1986).

[11]Y. Sargin, M. Kizilyalli, C. Telli, and H Guler, "A New Method for the Solid-State Synthesis of Tetracalcium Phosphate, a Dental Cement: X-Ray Powder Diffraction and IR Studies," *J. Eur. Ceram. Soc.*, **17** 963-970 (1997).

[12]K. Ishikawa, S. Takagi, L. C. Chow, and K. Suzuki, "Reaction of Calcium Phosphate Cements with Different Amounts of Tetracalcium Phosphate and Dicalcium Phosphate Anhydrous," *J. Biomed. Mater. Res.*, **46** 504-510 (1999).

[13]J.-J. Bian, D.-W. Kim, and K.-S. Hong, "Phase Transformation and Sintering Behavior of $Ca_2P_2O_7$," *Mater. Lett.*, **58** 347-351 (2004).

[14]W. E. Brown and L. C. Chow, "A New Calcium Phosphate Water-Setting Cement," in *Cement Research Progress*, P. W. Brown (Ed.), The American Ceramic Society, Westerville, Ohio, 1986, pp. 351-379.

[15]Norian® SRS® (Skeletal Repair System), Synthes Corp., Oberdorf, Switzerland.

[16]Biopex®, Mitsubishi Materials Corp., Tokyo, Japan

[17]A. C. Tas, "Synthesis of Biomimetic Ca-Hydroxyapatite Powders at 37°C in Synthetic Body Fluids," *Biomaterials*, **21** 1429-1438 (2000).

[18]K. S. TenHuisen and P. W. Brown, "Phase Evolution during the Formation of α-Tricalcium Phosphate," *J. Am. Ceram. Soc.*, **82** 2813-2818 (1999).

[19]C. Hamanishi, K. Kitamoto, K. Ohura, S. Tanaka, and Y. Doi, "Self-setting, Bioactive, and Biodegradable TTCP-DCPD Apatite Cement," *J. Biomed. Mater. Res.*, **32** 383-389 (1996).

[20]A. C. Tas, "A New Calcium Phosphate Cement Composition and a Method for the Preparation thereof," *European Patent No. EP 1,394,132*, March 2004.

[21]K. Ishikawa, S. Takagi, L. C. Chow, and Y. Ishikawa, "Properties and Mechanisms of Fast-setting Calcium Phosphate Cements," *J. Mater. Sci. Mater. M.*, **6** 528-533 (1995).

[22]U. Gbureck, J. E. Barralet, M. Hoffmann, and R. Thull, "Mechanical Activation of Tetracalcium Phosphate," *J. Am. Ceram. Soc.*, **87** 311-313 (2004).

[23]A. C. Tas, "Method of Preparing Alpha- and Beta-Tricalcium Phosphate Powders," *US Patent Appl. No. 2003/0235622*.

[24]U. Posset, E. Loecklin, R. Thull, and W. Kiefer, "Vibrational Spectroscopic Study of Tetracalcium Phosphate in Pure Polycrystalline Form and As constituent of a Self-setting Bone Cement," *J. Biomed. Mater. Res.*, **40** 640-645 (1998).

[25]T. Kokubo, "Apatite Formation on Surfaces of Ceramics, Metals and Polymers in Body Environment," *Acta Mater.*, **46** 2519-2527 (1998).

[26]X. Yin and M. J. Stott, "Biological Calcium Phosphates and Posner's Cluster," *J. Chem. Phys.*, **118** 3717-3723 (2003).

[27]A. S. Posner and F. Betts, "Synthetic Amorphous Calcium Phosphate and its Relation to Bone Mineral Structure," *Acc. Chem. Res.*, **8** 273-281 (1975).

[28]F. Barrere C. A. van Blitterswijk, K. de Groot, and P. Layrolle, "Influence of Ionic Strength and Carbonate on the Ca-P Coating Formation from SBF x 5 Solution," *Biomaterials*, **23** 1921–1930 (2002).

[29]K. Onuma and A. Ito, "Cluster Growth Model for Hydroxyapatite," *Chem. Mater.*, **10** 3346-3351 (1998).

[30]S. R. Radin and P. Ducheyne, "The Effect of Calcium Phosphate Ceramic Composition and Structure on *In Vitro* Behavior. II. Precipitation," *J. Biomed. Mater. Res.*, **27** 35-45 (1993).

[31]S. Matsuya, S. Yakagi, and L. C. Chow, "Hydrolysis of Tetracalcium Phosphate in H_3PO_4 and KH_2PO_4," *J. Mater. Sci.*, **31** 3263-3269 (1996).

[32]W.-C. Chen, J.-H. Chern Lin, and C.-P. Ju, "Transmission Electron Microscopic Study on Setting Mechanism of Tetracalcium Phosphate/Dicalcium Phosphate Anhydrous-based Calcium Phosphate Cement," *J. Biomed. Mater. Res.*, **64A** 664-671 (2003).

CALCIUM PHOSPHATE BASED CERAMICS VIA SPINODAL DECMOPOSITION

Michael Sansoucie and Robert W. Hyers
University of Massachusetts
Mechanical & Industrial Engineering
Amherst, MA 01376

ABSTRACT

Formation of hydroxyapatite through spinodal decomposition is analyzed. Spinodal decomposition involves a thermodynamic separation of a single-phase material into an interconnected, two-phase material with a continuous network. The $CaO-P_2O_5-FeO_n$ ternary system used in this research contains a liquid miscibility gap and undergoes spinodal decomposition. For this research the primary phase is a mixture of hydroxyapatite (HAp, $Ca_{10}(PO_4)_6(OH)_2$) and tricalcium phosphate (TCP, $Ca_3(PO_4)_2$), and the secondary phase is iron oxide.

INTRODUCTION

Bone diseases and defects are one of the most common medical conditions that have a direct impact on the quality of life[1]. Furthermore, bone is the second most implanted material in humans after blood. There are over 450,000 bone grafts in the United States and over 2.2 million worldwide[2]. This accounts for billions of dollars in market potential.

Due to the increasing lifespan of humans, an implanted material must have a very long service life, typically up to 20 years or more[3]. This makes it difficult to use bioinert materials such as metals and nonsorbable ceramics, because they can break down from wear and are prone to fracture due to fatigue. Furthermore, the wear products cause inflammation, potentially leading to arthritis or cancer.

A disadvantage of using metals and ceramics as bone implants is stress shielding, which was discovered by Julius Wolff in 1892. Natural living bone adapts to the load it experiences. Materials implanted into bone share the load that the bone would normally see. If the elastic modulus of the implant material is much higher than the elastic modulus of the natural bone, the bone will experience proportionally lower stresses. This will cause the bone to adapt to this lower stress by atrophy[4], making most nonresorbable materials which have much higher elastic moduli, a poor choice for synthetic bone.

A resorbable biomaterial circumvents this issue, because it will be gradually broken down and replaced with new natural bone through normal bone remodeling[3]. This would give the patient enough time for complete recovery while supporting the patient initially. The amount of time it takes for this to occur can be changed by altering the chemical composition of the synthetic bone[5]. This resorbtion rate can also be tailored to the severity of the injury. If the injury is fairly minor, then the implant can be made specifically to be resorbed much quicker, and vice versa.

Natural bone consists mainly of minerals and collagen fibers, with approximately 69% by weight being hydroxyapatite (HAp, $Ca_{10}(PO_4)_6(OH)_2$)[6]. Bioceramics have the advantage of being very similar in chemistry and structure to that of natural bone. The most widely used bioceramics are hydroxyapatite and tricalcium phosphate[7]. Previous research has shown that natural bone grows into porous, synthetic HAp producing a strong bond[8-11]. The synthetic bone provides support and shape until it is replaced by the body with natural bone. It has been shown that 37% of pore volumes must be enclosed by interconnections greater than 100μm for mineralized bone ingrowth[12].

Porosity and interconnectivity determine the amount and type of ingrowth. For implants with a high degree of porosity and interconnectivity, tissue ingrowth can begin as early as 3 or 4 days and may be complete by 28 days[13]. This also depends on many factors including surface characteristics, composition, stress, location of implant, etc.

Resorbable bioceramics contain only elements that can easily be processed through normal metabolic means, such as calcium and phosphorus. Both hydroxyapatite and tricalcium phosphate are resorbable bioceramics[14], but hydroxyapatite dissolves much more slowly than TCP in a variety of fluids. It has been shown that when dense HAp and TCP of similar purity and microstructure were compared, the TCP dissolved 12.3 times faster in acid solutions (pH 5.2) and 22.3 times faster in basic solutions (pH 8.2)[6].

Current methods for the production of hydroxyapatite typically involve sintering at high temperatures[6]. However, sintering at high temperatures causes low porosity, closed pores, and poor phase purity (due to decomposition), while sintering at lower temperatures to increase porosity produces weak bonds[15]. Porous sintered ceramics are currently used in alveolar ridge augmentation, maxillofacial reconstruction, and implant coatings[16], but sintering is an inadequate method for producing good quality HAp implant material for structural bone replacement.

An alternative method for the production of hydroxyapatite bioceramics is spinodal decomposition. Liquids are not always completely soluble in one another. This causes the liquids to separate into two phases and is called liquid immiscibility[17]. The $CaO-P_2O_5-FeO_n$ ternary system, a phase diagram of which is shown in Figure 1, contains a liquid miscibility gap. The primary phase will be a mixture of hydroxyapatite (HAp, $Ca_{10}(PO_4)_6(OH)_2$) and tricalcium phosphate (TCP, $Ca_3(PO_4)_2$), both of which are calcium phosphates. The secondary phase is iron oxide in this case. Tricalcium phosphate resorbs faster than HAp; therefore, the resorbtion rate may be controlled by altering the HAp/TCP ratio through solution treatment. Control of these process variables would allow the implant material to be tailored to the specific application. Different bones and regions of bones have different porosities and pore sizes; therefore, control of these characteristics is very favorable.

The proposed method for the production of calcium phosphate implant materials should allow the implant to be tailored to the specific application. The resorption rate, porosity, and pore size can all be modified. The wide variations in natural bone, makes an implant that can be customized to each and every specific application very appealing.

For example, a slowly resorbable calcium phosphate ceramic should not be used in situations where bone remodeling, bone replacement, or original strength of bone is desired. Highly resorbable calcium phosphate ceramics should be chosen in situations where resorption and associated bone replacement is desired because calcium phosphate ceramics allow greater remodeling, vascularity, and restoration of strength with time [14].

If the iron oxide is dissolved, this would leave a porous, continuously-connected structure of HAp and TCP.

METHODOLOGY

In order to create the samples, the correct mixture of compounds had to first be generated. The phase diagram served as a guide for choosing the compositions to be analyzed. At first, compositions towards the middle of the immiscibility region were tested.

The chemicals that were available for the generation of samples consisted of iron oxide Fe_2O_3, Alfa Aesar), calcium oxide (CaO, Acros Organics), and calcium phosphate ($Ca_2P_2O_7$, Alfa Aesar). In order to be sure that the correct composition on the phase diagram was achieved, the correct stoichiometries had to be calculated.

By consulting the Ellingham diagram, it was determined that Fe_3O_4 would be the stable polymorph of iron oxide at the temperatures used.

The calcium phosphate decomposes to calcium oxide with a stoichiometry similar to the following:

$$Ca_2P_2O_7 \rightarrow 2CaO + P_2O_5.$$

The iron oxide transforms as shown:

$$6Fe_2O_3 \rightarrow 4Fe_3O_4 + 3O_2.$$

Calcium oxide had to be added to obtain the correct mixtures.

Ten gram samples were made by using the weight percentages from the phase diagram. The starting powders were mixed well and melted in a platinum crucible in a Deltech DT31 vertical tube furnace (Deltech Inc., Denver, CO), which was modified for these experiments. The molten material was poured onto a steel plate and allowed to cool.

After cooling, the sample materials were sectioned using a Buehler ISOMET® Low Speed Saw. The sectioned samples were then set in Struers Epofix epoxy and allowed to harden overnight. Next, the mounted samples were polished using standard ceramographic techniques. No etching was necessary to see the microstructure with either an optical microscope or a scanning electron microscope.

The microstructure was typically viewed first using an Olympus Vanox-T optical microscope. Pictures were taken from this microscope with a Kodak Digital Science DC40 camera. Samples were then often analyzed with a FEI Quanta 200 Environmental Scanning Electron Microscope (ESEM).

X-ray diffraction analysis was performed on several samples to determine the phases present. For this analysis, the spinodal decomposed region of the sample was crushed into a powder using a porcelain mortar and pestle. The powder was then fixed to a piece of clay that was forced through the hole of the sample holder. This allowed the sample to sit flush with the face of the sample holder.

RESULTS AND DISCUSSION

Figure 1 shows the microstructures seen from various locations on the phase diagram. From the phase diagram, it is expected that the separated phases are nearly pure iron oxide and a mixture of calcium phosphates and phosphores.

Figure 2 shows X-ray diffraction results from one sample. Other samples show similar results, which demonstrate reproducibility. Not all of the peaks were matched; however, several hydroxyapatite (HAp) peaks do appear. A tricalcium phosphate (TCP) and iron oxide (Fe_3O_4) peak was also identified. This indicates that the primary phase is biphasic calcium phosphate with a high HAp/TCP ratio.

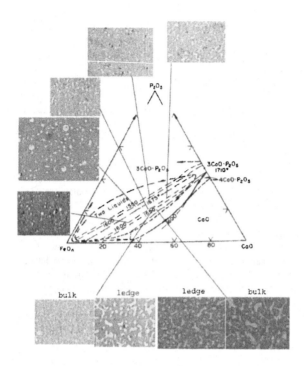

Figure 1: Microstructure results shown on the phase diagram. Phase diagram reprinted with permission of The American Ceramic Society, www.ceramics.org. Copyright 1984. All rights reserved.

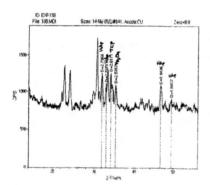

Figure 2: X-ray diffraction results.

Several samples show spinodal decomposition phase separation, and some samples show the initial stages of spinodal phase separation. Unfortunately, most samples produce macrosegregation such as that shown in Figure 3.

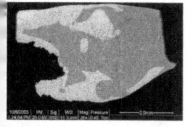

Figure 3: Micrograph showing macrosegregation. The light phase is mostly iron oxide, and the dark phase is primarily calcium phosphate (HAp with some TCP).

Some samples show a transition in the wavelength of the spinodal decomposition as shown in Figure 4. This micrograph was taken towards the bottom of the sample, which would receive the most pronounced temperature gradient due to the quench plate. The smaller wavelength microstructure is due to a faster cooling rate, and the longer wavelength is due to a slower cooling rate.

Figure 4: Micrograph shows a transition in the wavelength of the spinodal decomposition from fine near the quench plate (bottom) to coarse in the bulk (top). The light phase is primarily iron oxide, and the dark phase is mostly calcium phosphate (HAp with some TCP).

Some regions show uniform spinodal decomposition separation with consistent wavelength and well-connected structure, such as that shown in Figure 5.

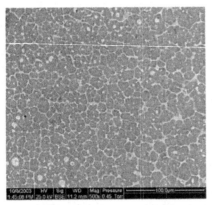

Figure 5: Micrograph showing spinodal decomposition. However, this sample also contains macrosegregation. The light phase is mostly iron oxide, and the dark phase is primarily calcium phosphate (HAp with some TCP).

Heat treating the samples produced no changes in the microstructure. This indicates that the structure is fully developed.

CONCLUSIONS

Formation of hydroxyapatite through spinodal decomposition was analyzed, and it has been shown that a spinodal region does occur in the $CaO-P_2O_5-FeO_n$ ternary system and that it does generate microstructures typical of spinodal decomposition.

Some samples showed a transition in the wavelength of the spinodal decomposition. This is due to temperature gradients when the sample was quenched. The smallest wavelength microstructure is seen towards the bottom of the sample, which would have the fastest cooling rate due to the quench plate. The larger wavelength microstructure is seen in regions where the cooling rate would be slower. This demonstrates that the porosity of the sample can be controlled through thermal history.

The X-ray diffraction results show that a biphasic calcium phosphate material can be created from this method.

REFERENCES

[1]H.-M. Kim, "Bioactive Ceramics: Challenges and Perspectives," *Journal of the Ceramic Society of Japan*, **109** [4] S49-S57 (2001).

[2]J.D. Gresser, D.J. Trantolo, D.L. Wise, and K.-U. Lewandrowski, "Engineering of Resorbable Grafts for Craniofacial Reconstruction"; pp. 675-697 in *Biomaterials and Bioengineering Handbook*. Edited by D. Wise. Marcel Dekker, New York, 2000.

[3]V. Dubok, "Bioceramics - Yesterday, Today, Tomorrow," *Powder Metallurgy and Metal Ceramics*, **39** [7-8] 381-394 (2000).

[4]C.M. Agrawal, "Reconstructing the Human Body Using Biomaterials," *JOM*, 31-35 (1998).

[5]N. Kivrak and A.C. Tas, "Synthesis of Calcium Hydroxyapatite-Tricalcium Phosphate (HA-TCP) Composite Bioceramic Powders and Their Sintering Behavior," *Journal of the American Ceramic Society*, **81** [9] 2245-2252 (1998).

[6]M. Jarcho, "Biomaterial Aspects of Calcium Phosphates," *Dental Clinics of North America*, **30** [1] 25-47 (1986).

[7]M. Jarcho, "Calcium Phosphate as Hard Tissue Prosthetics," *Clinical Orthopaedics and Related Research*, **157** 259-278 (1981).

[8]R.E. Holmes and S.M. Roser, "Porous hydroxyapatite as a bone graft substitute in alveolar ridge augmentation: a histometric study," *International Journal of Oral and Maxillofacial Surgery*, **16** 718-728 (1987).

[9]R.V. Rejda, J. Peelen, and K. deGroot, "Tri-Calcium Phosphate as a Bone Substitute," *Journal of Bioengineering*, **1** 93-97 (1977).

[10]R.Z. LeGeros, "Biodegradation and Bioresorption of Calcium Phosphate Ceramics," *Clinical Materials*, **14** 65-88 (1993).

[11]H. Schliephake, F. Neukam, and D. Klosa, "Influence of pore dimensions on bone ingrowth into porous hydroxylapatite blocks used as bone graft substitutes," *International Journal of Oral and Maxillofacial Surgery*, **20** 53-58 (1991).

[12]J. Klawitter and S. Hulbert, "Application of Porous Ceramics for the Attachment of Load Bearing Internal Orthopedic Applications," *Journal of biomedical Materials Research Symposium*, **2** [1] 161-229 (1971).

[13]L.L. Hench and J. Wilson, editors. *An Introduction to Bioceramics*. Singapore: World Scientitic; 1993.

[14]R. LeGeros, J. Parsons, S. Liu, S. Metsger, D. Peterson, and M. Walker, "Significance of the Porosity and Physical Chemistry of Calcium Phosphate Ceramics: Biodegradation-Bioresorption," *Annals of the New York Academy of Sciences*, **523** 268-272 (1988).

[15]P. Brown, R. Martin, and K. TenHuisen, "Factors Influencing the Formation of Monolithic Hydroxyapatite at Physiological Temperature"; pp. 37-48 in *Bioceramics: Materials and Applications II (Ceramic Transactions v63)*. Edited by R. Rusin, G. Fischman. American Ceramic Society, Westerville, OH, 1995.

[16]R. LeGeros and J. LeGeros, "Calcium Phophate Bioceramics: Past, Present and Future"; pp. 3-10 in *Bioceramics (Key Engineering Materials Vol. 240-242)*. Edited by B. Ben-Nissan, D. Sher, W. Walsh. Trans Tech Publications, Switzerland, 2003.

[17]C.G. Bergeron and S.H. Risbud, *Introduction to Phase Equilibrium in Ceramics*, Columbus: The American Ceramic Society, 1984.

PREPARATION OF BRUSHITE POWDERS AND THEIR *IN VITRO* CONVERSION TO NANOAPATITES

A. Cuneyt Tas and Sarit B. Bhaduri
School of Materials Science and Engineering, Clemson University, Clemson, SC 29634, USA

ABSTRACT

Brushite (DCPD: dicalcium phosphate dihydrate: $CaHPO_4 \cdot 2H_2O$) powders were chemically synthesized by using Na- and K-phosphate and calcium chloride-containing aqueous solutions at room temperature (RT), followed by drying at 37°C. DCPD powders thus formed were found to contain 460 ppm K and 945 ppm Na. Upon calcining in air these powders readily transformed into $CaHPO_4$ (monetite) first, and then into $Ca_2P_2O_7$. Na- and K-doped DCPD powders were shown to completely transform, in less than 1 week, into poorly crystalline carbonated apatite upon immersion in an acellular simulated body fluid (SBF) solution at 37°C. The tris-buffered SBF solution used in this study had a carbonate ion concentration of 27 mM, which was equal to that of human blood plasma. This finding suggests the use of these DCPD powders as potential bone-substitute materials, which can be easily manufactured in aqueous solutions friendly to living tissues at temperatures between RT and 37°C.

INTRODUCTION

Potential bone substitute materials must be actively resorbed [1] *in vivo* by the osteoclasts (cells that are able to resorb the fully mineralized bone), as they are equipped with a variety of enzymes, which lower the local pH to a range of 3.9 to 4.2. This occurs via a process called cell-mediated acidification in which the host bone can deposit new bone on those resorption sites by the osteoblasts (cells that build the extracellular matrix and regulate its mineralization). Bulk ceramics in the form of porous prismatic blocks, self-hardening cements, granules, coatings obtained by the high-temperature thermal spray techniques, injectable pastes, etc. are used as implant materials. For their successful application, they should be able to fully take part in the bone remodeling processes and must be eventually resorbed and fully replaced by the new bone within a year following the implantation [2]. Either synthetic or bovine calcium hydroxyapatite ($Ca_{10}(PO_4)_6(OH)_2$) bioceramics heated at or above 1100°C during their processing do not resorb well and do not take part in the *in vivo* bone remodeling processes within the aforementioned time frame [3, 4]. In general, resorbability of highly crystalline, sintered bovine-origin apatitic calcium phosphates is poor, due to the volatilization of initially present HPO_4^{2-} and CO_3^{2-} ions.

DCPD (dicalcium hydrogen phosphate dihydrate) can be synthesized in aqueous solutions at room temperature by using water-soluble calcium (e.g., $Ca(NO_3)_2 \cdot 4H_2O$, $CaCl_2 \cdot 2H_2O$, or $Ca(CH_3COO)_2 \cdot H_2O$) and phosphate (e.g., $NH_4H_2PO_4$, $(NH_4)_2HPO_4$, Na_2HPO_4, NaH_2PO_4, KH_2PO_4 or K_2HPO_4) salts upon adjusting the Ca/P molar ratio to 1 [5-8]. The use of Na- and/or K-containing starting chemicals during the synthesis may also result (although it may not be necessarily and strictly so) in the production of Na- and/or K-doped DCPD powders [9-12]. On the other hand, pure DCPD powders can also be synthesized, for instance, by reacting a suspension of $Ca(OH)_2$ with stoichiometric amounts of H_3PO_4, as long as the solution pH is kept in the acidic range [13, 14]. DCPD will transform (by losing its crystal water) into $CaHPO_4$ (DCPA: monetite) upon heating at or above 110°C [15]. Both DCPD [16, 17] and DCPA [18] phases are successfully used as starting materials in the preparation of the powder components of self-hardening apatitic calcium phosphate cements. However, recent reports [19, 20] are directed towards the development of self-hardening orthopedic cements whose final product is DCPD. Such formulations have superior *in vivo* resorbability, as opposed to conventional apatitic cements. This scientific basis behind this development can be clearly explained by the dissolution data published by Tang *et al.* [21]. Relying on the experimental solubility values of some of the calcium phosphate phases recently reported by Tang *et al.* [21], it is seen that DCPD has a dissolution rate of 4.26×10^{-4} mol/m^{-2} min^{-1} at a pH value of 5.5, and this rate is only about 3.4 times greater than that of $Ca_3(PO_4)_2$ (i.e., 1.26×10^{-4}). To compare these, the dissolution rate for carbonated apatite was reported by the same researchers [21] to be 1.42×10^{-6}.

Kumar *et al.* [10, 11] reported previously that electrodeposited DCPD, which was doped with monovalent cations, such as potassium, might result in more rapid transformation into apatitic calcium phosphates upon soaking the samples in Hanks Balanced Salt Solution (HBSS) [22]. HBSS is the historical origin of Simulated Body Fluid (SBF) solutions, which were popularized by Kokubo [23] over the last decade. The main difference between an HBSS solution and SBF lies in the value of the Ca/P molar ratios. HBSS has a Ca/P ratio of 1.823, whereas the same in SBF is 2.50. Owing to its lower Ca/P molar ratio, an HBSS solution, in contrast to SBF, needs quite long times to induce apatitic calcium phosphate formation [24]. However, to raise the Ca/P ratio of HBSS to 2.50, one only needs to add 70.5 mg of $CaCl_2 \cdot 2H_2O$ into 1 liter of a commercially available HBSS solution. Therefore, soaking DCPD powders in SBF, instead of HBSS solutions, to test their apatite inducing ability would be more viable. *Tas*-SBF [25] used in this study was a Tris-HCl buffered solution with a HCO_3^- ion concentration equal to 27 mM, whose preparation details were previously explained elsewhere.

The main purpose of this study was to develop a robust chemical synthesis procedure for the manufacture of Na- and K-doped DCPD powders, and then study their transformation into apatite [24] by immersing them in an acellular and metastable (with respect to hydroxyapatite nucleation) SBF solution [25] over the duration of 36 hours to 5 weeks. High-temperature calcination behavior of the alkali-doped DCPD powders was also studied and reported.

EXPERIMENTAL PROCEDURE
The synthesis procedure used to form Na- and K-doped DCPD powders in this study consisted of preparing two solutions. Solution-A is prepared as follows; 4.127 g KH_2PO_4 was dissolved in 3.5 L of deionized water, followed by the addition of 15.065 g Na_2HPO_4, which resulted in a solution of pH 7.4 at RT ($22\pm2°C$). Solution-B (of pH 7.3) was simply prepared by dissolving 20.068 g of $CaCl_2 \cdot 2H_2O$ in 250 mL of deionized water. Solution-B was then added at once into solution-A, and the precipitates that formed were aged for 80 minutes at RT under continuous but moderate stirring (final solution pH = 5.3). Solids recovered from their mother liquors were dried for 2 days at 37°C in an air atmosphere to obtain 16.85 g of Na- and K-doped DCPD powders. The percentage yield of the powder synthesis process was $72\pm1\%$.

Calcination behavior of these powders were determined by isothermal heatings over the temperature range of 300° to 1000°C, with 6 h of soak time at the peak temperatures. Simultaneous TG/DTA runs (RT to 1000°C) were also performed on the DCPD samples.

Powder samples were characterized, at all stages, by XRD, SEM, EDXS, FTIR, ICP-AES, and TG/DTA analysis.

Hydrolytic conversion of Na- and K-doped DCPD into poorly crystalline, carbonated, apatitic calcium phosphate powders were studied by soaking in a *Tas*-SBF solution [25] at 37°C. 250 mg portions of DCPD powders were placed in 25 mL of the *Tas*-SBF [25] solution (2.5 mM Ca^{2+}, 1 mM HPO_4^{2-}, 27 mM HCO_3^-, 142 mM Na^+, 5 mM K^+, 1.5 mM Mg^{2+}, 0.5 mM SO_4^{2-}, 125 mM Cl^-, *tris*-buffered, pH=7.4), in plastic vials. During the conversion process, 15 mL aliquots of solutions were replenished with a fresh SBF solution at every 36 hours. The experiment continued in 8 identical vials for solid sample recovery times of 36 h, 72 h, 1 week, 1.5 weeks, 2 weeks, 3 weeks, 4 weeks and 5 weeks. The vial contents were then filtered and washed with 400 mL of deionized water and dried at 37°C for 48 hours, prior to characterization runs.

RESULTS AND DISCUSSION
Chemically-precipitated powders were characterized by XRD, FTIR, SEM, ICP-AES and TG/DTA data to be single-phase, Na- and K-doped (460 and 945 ppm, respectively) $CaHPO_4 \cdot 2H_2O$, as shown in Figures 1a (the inset displays the FTIR data) through 1c. Brushite crystallizes in the monoclinic space group Cc with the lattice parameters, a=6.359, b=15.177, c=5.81Å, β=118.54° [26]. As a function of increasing calcination temperature, as shown in Figures 1d to 1e (XRD and FTIR data, respectively) brushite first transforms into $CaHPO_4$ (DCPA, monetite) and then to $Ca_2P_2O_7$. DCPA has the following lattice parameters; a=6.910, b=6.627, c=6.998 Å, α=96.34°, β=103.82°, and γ=88.33°. Its structure consists of $CaHPO_4$ chains bonded together by Ca-O bonds and three types of hydrogen bonds [26]. The plate-like morphology of optically transparent brushite crystals (Fig. 1b) was preserved even after conversion into microporous $Ca_2P_2O_7$ by heating in air at 1000°C for 6 hours (Fig. 1e).

Calcination behavior of plate-like DCPD powders (Figs. 1b and 1e) of this study presented a typical case for the thermally induced transformation of an orthophosphate into a pyrophosphate according to the reactions:

$$2CaHPO_4 \cdot 2H_2O \text{ (s)} \rightarrow 2CaHPO_4 \text{ (s)} + 4H_2O \text{ (g)} \qquad (1)$$
$$2CaHPO_4 \text{ (s)} \rightarrow Ca_2P_2O_7 \text{ (s)} + H_2O \text{ (g)} \qquad (2).$$

These reactions were experimentally confirmed to take place by the TG/DTA analysis (Fig. 1c). β-$Ca_2P_2O_7$ phase obtained after 850° and 1000°C calcinations conformed to the ICDD PDF 9-346. On the other hand, α–$Ca_2P_2O_7$ phase (ICDD PDF 9-345) was observed in the samples calcined at 500° and 700°C (Fig. 1d). TG/DTA data of DCPD powders (i.e., Fig. 1c) also agreed well with the work of Joshi *et al* [27]. We observed that the brushite transformed to monetite at around 180°C with a weight loss of 20.3%, and the monetite to $Ca_2P_2O_7$ transition was completed above 440°C, with a further 6% weight loss. FTIR data presented in Figures 1a (as-formed DCPD) and 1f (as a function of calcination temperature) also coincided very well with those in literature [27-29].

Lee *et al*. [30] have recently tested commercial powders of $Ca_2P_2O_7$ as a synthetic bone graft material in comparison to hydroxyapatite. They reported a superior resorbability for $Ca_2P_2O_7$ in their canine-based proximal tibia model. The authors concluded that the calcium pyrophosphate initially functioned as an osteoconductive scaffolding, and within three months it consecutively and seamlessly took part in the bone remodeling process.

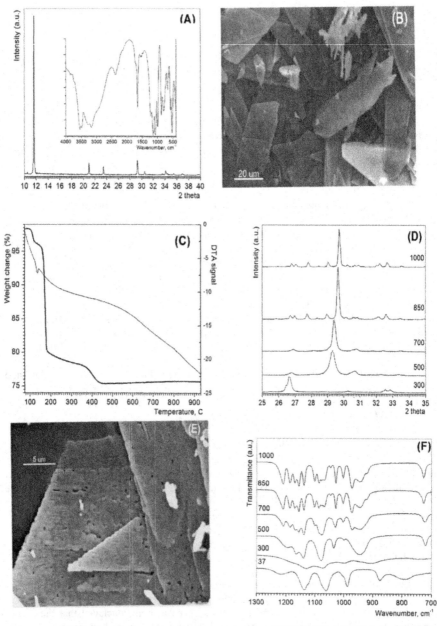

Fig. 1. *(A to C)*: XRD, FTIR, SEM, TG characterization of DCPD, *(D to F)*: Calcined DCPD

DCPD is known to be a nucleation precursor, in aqueous solutions, to the apatitic calcium phosphates [5]. DCPD transforms into the thermodynamically more stable, apatitic calcium phosphate, by a dissolution-reprecipitation mechanism [31]. DCPD has a relatively low solubility in water, and thus water alone could not be sufficient to drive the reprecipitation mechanism [32]. However, if the aqueous medium of DCPD immersion contains Ca^{2+} ions then the process will readily proceed according to the following reaction [33]:

$$6CaHPO_4 \cdot 2H_2O + (4-x)CaCl_2 + (2-x)H_2O \rightarrow$$
$$Ca_{10-x}(HPO_4)_x(PO_4)_{6-x}(OH)_{2-x} + (8-2x)HCl + 12H_2O \qquad (3).$$

Apatite formed in reaction (3) is termed as Ca-deficient hydroxyapatite (CDHA), and this formula represents the family of apatites formed at a neutral pH. However, this formula is still a simplified version since it does not define the carbonate ions incorporated into the structure [34].

Therefore, we selected a tris-buffered, carbonated (27 mM HCO_3^-) SBF solution [25] of pH 7.4 as the immersion medium to examine the dissolution-reprecipitation mechanism of our DCPD powders as a function of time. If we were to soak the DCPD powders in pure water [11, 33], we would have mostly seen its sluggish dissolution over a period of 1 month. As seen in the XRD data of Fig. 2a for the DCPD powders immersed in SBF solutions at 37°C, even after 72 hours of soaking a significant amount of CDHA was formed as predicted by reaction (3). By the end of 1 week of soaking time, all the crystal peaks of DCPD disappeared from the XRD patterns. FTIR data given in Fig. 2b for those samples clearly indicated the same trend. The recognition of the characteristic bands in the FTIR spectra of apatite and DCPD phases have been unequivocally established in various references [25-29]. FTIR spectra of Fig. 2b displayed that the carbonated nature of the apatitic calcium phosphate initially formed was gradually developing with an increase in soaking time. SEM photomicrographs given in Figs. 2c and 2d, as a function of immersion time in SBF, depicted the characteristic morphology of needle-like CDHA.

From X-ray or neutron crystallographic studies, crystal structure of DCPD is known to contain compact sheets or bilayers parallel to the (010) plane [26]. One bilayer was found to present sheets of calcium and phosphate ions, while the other bilayer comprised water molecules. Flade et al. [35] reported that the hydrated bilayer was the terminating layer at the surface of the (010) face in aqueous solutions, and dissolution of DCPD must start from the ledges. To examine the dissolution behavior of DCPD more explicitly, we performed a slightly modified experiment. In that experiment, we placed a 30 mL aliquot of freshly prepared Solution-A (its preparation was described in the Experimental Section) into a 50 mL-capacity glass beaker covered with Parafilm®. Beaker was then heated to 37°C in a constant temperature oven. 1 minute after the injection (with a syringe and needle) of 1 mL of Solution-B into that beaker, solution pH was recorded as 5.5 at 37°C, and it was rapidly cooled to 0°C by immersing it into an ice-water bath. The precipitates were separated immediately from the mother liquor by centrifugal filtration, and washed with water. The formed precipitates as shown in the optical micrograph given in Figure 3a had a star- or rosette-like morphology. They were shown by XRD to be highly crystalline, single-phase DCPD. These crystals were regarded as the early-stage crystallization products of the synthesis process reported in this study.

To picture the dissolution behavior of those DCPD crystals, a 100 mg portion was placed in 10 mL of a freshly prepared Tas-SBF solution [25]. Following 3 hours of immersion at 37°C in SBF, the DCPD crystals were washed with water and dried at 37°C, overnight. XRD pattern of the samples again indicated single-phase DCPD, as shown in Figure 1a. SEM photomicrograph given in Figure 3b showed the dissolution of DCPD, revealing the aforementioned bilayer structure, which until now could only be ascertained by using crystallographic techniques. Reprecipitation leg of this mechanism, which leads to the apatitic calcium phosphate formation, subsequently takes place at the later stages of SBF immersion, as was shown above. SBF solutions of pH 7.2 to 7.4 are supersaturated with respect to apatite (i.e., Ca/P molar ratio = 2.50). Therefore, the only phase that can precipitate from such neutral pH solutions is carbonated hydroxyapatite.

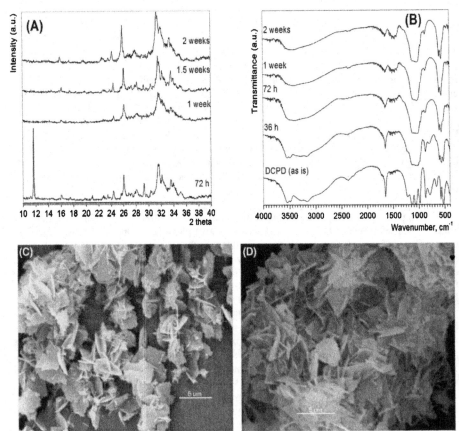

Fig. 2 (A-B): XRD and FTIR data of DCPD powders soaked in SBF at 37°C, (C): 36 h of immersion in SBF, (D): 1 week in SBF; apatitic needles visible in both SEM pics of (C) and (D)

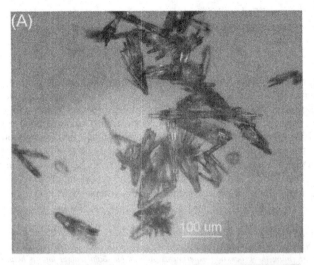

Fig. 3. (a) Early-stage crystals of DCPD formed in 1 minute, (b) dissolution of DCPD crystals in SBF at 37°C, revealing the bilayer structure

Conversion of $CaHPO_4 \cdot 2H_2O$ into poorly crystalline apatite becomes important when its possible uses as synthetic bone substitutes or bone defect filling materials are considered. Self-setting orthopedic cements based on DCPD have already been formulated and reported for their enhanced resorbability [20]. Moreover, the plate- or needle-like morphology observed in chemically prepared DCPDs might have possible uses in improving the mechanical properties (e.g., compressive and flexural strength) of the self-setting orthopedic cements, which include DCPD as a constituent [36]. Doping of DCPD powders with small amounts of Na and K during their synthesis, as exemplified in this study, imparted a neutral surface pH (i.e., 6.90-7.10) to those, in comparison to pure, commercially available DCPD powders.

CONCLUSIONS
(1) The chemical process outlined here allowed the uncomplicated synthesis of biocompatible Na- and K-doped DCPD powders, with a unique plate-like morphology, by starting with aqueous solutions at the physiologic pH and temperature conditions.
(2) The results showed that it is possible to preserve this particle morphology even though the powders were later converted to $Ca_2P_2O_7$ by high-temperature calcination.
(3) Finally, this study also exemplified a robust procedure for producing carbonated apatitic calcium phosphate powders, after simple immersion of DCPD powders in SBF solutions (pH=7.4) at 37°C. The highest temperature of processing hereby employed in the manufacture of bulk, poorly crystalline apatite powders was 37°C.

ACKNOWLEDGMENTS

This work was partially supported by NSF DMII 0085100. Authors are also grateful for the technical help of Mr. Baris Kokuoz in performing the processing experiments, and to Mr. Sahil Jalota for his help with some of the characterization work.

REFERENCES
[1]F. Monchau, A. Lefevre, M. Descamps, A. Belquin-Myrdycz, P. Laffargue, and H.F. Hildebrand, "In Vitro Studies of Human and Rat Osteoclast Activity on Hydroxyapatite, β-Tricalcium Phosphate, Calcium Carbonate," *Biomolecular Engineering*, 19, 143-152 (2002).
[2]R. Gunzburg, M. Szpalski, N. Passuti and M. Aebi, *The Use of Bone Substitutes in Spine Surgery*, pp. 2-11, Springer-Verlag, Berlin, 2002.
[3]S. Joschek, B. Nies, R. Krotz, A. Gopferich, "Chemical and Physicochemical Characterization of Porous Hydroxyapatite Ceramics Made of Natural Bone," *Biomaterials*, 21, 1645-1658 (2000).
[4]P. Ducheyne, "Bioceramics: Material Characteristics versus In Vivo Behavior," *Journal of Biomedical Materials Research: Applied Biomaterials*, 21, 219-236 (1987).
[5]P.A. Ngankam, P. Schaaf, J.C. Voegel, and F.J.G. Cuisinier, "Heterogeneous Nucleation of Calcium Phosphate Salts at a Solid/Liquid Interface Examined by Scanning Angle Reflectometry," *Journal of Crystal Growth*, 197, 927-938 (1999).
[6]G.R. Sivakumar, E.K. Girija, S. Narayana Kalkura, C. Subramanian, "Crystallization and Characterization of Calcium Phosphates: Brushite and Monetite," *Cryst. Res. Tech.*, 33, 197-205 (1998).
[7]J.S. Sorensen and H.E. Lundager Madsen, "The Influence of Magnetism on Precipitation of Calcium Phosphate," *Journal of Crystal Growth*, 216, 399-406 (2000).
[8]J. Xie, C. Riley, M. Kumar, and K. Chittur, "FTIR/ATR Study of Protein Adsorption and Brushite Transformation to Hydroxyapatite," *Biomaterials*, 23, 3609-3616 (2002).
[9]R.P. Shellis, A.R. Lee, and R.M. Wilson, "Observations on the Apparent Solubility of Carbonate-Apatites," *Journal of Colloid and Interface Science*, 218, 351-358 (1999).

[10]M. Kumar, H. Dasarathy, and C. Riley, "Electrodeposition of Brushite Coatings and Their Transformation to Hydroxyapatite in Aqueous Solutions," *J. Biomed. Mater. Res.*, **45**, 302-310 (1999).

[11]M. Kumar, J. Xie, K. Chittur, and C. Riley, "Transformation of Modified Brushite to Hydroxyapatite in Aqueous Solution: Effect of Potassium Substitution," *Biomaterials*, **20**, 1389-1399 (2000).

[12]J. Redepenning, T. Schlessinger, S. Burnham, L. Lippiello, and J. Miyano, "Characterization of Electrolytically Prepared Brushite and Hydroxyapatite Coatings on Orthopedic Alloys," *Journal of Biomedical Materials Research*, **30**, 287-294 (1996).

[13]R.I. Martin and P.W. Brown, "Phase Equilibria Among Acid Calcium Phosphates," *Journal of The American Ceramic Society*, **80**, 1263-1266 (1997).

[14]A. Ferreira, C. Oliveira, and F. Rocha, "The Different Phases in the Precipitation of Dicalcium Hydrogen Phosphate Dihydrate," *Journal of Crystal Growth*, **252**, 599-611 (2003).

[15]*Handbook of Chemistry and Physics*, p. 4-49, 72nd ed. Edited by D.R. Lide. CRC Press, Boston, 1992.

[16]K. Kurashina, H. Kurita, M. Hirano, A. Kotani, C.P.A.T. Klein, and K. de Groot, "In Vivo Study of Calcium Phosphate Cements: Implantation of an α-Tricalcium Phosphate/Dicalcium Phosphate Dibasic/Tetracalcium Phosphate Monoxide Cement Paste," *Biomaterials*, **18**, 539-543 (1997).

[17]D. Knaack, M.E.P. Goad, M. Aiolova, C. Rey, A. Tofighi, and D.D. Lee, "Resorbable Calcium Phosphate Bone Substitute," *J. Biomed. Mat. Res. Appl. Biomat.*, **43**, 399-409 (1998).

[18]E.M. Ooms, J.G.C. Wolke, M.T. van de Heuvel, B. Jeschke, and J.A. Jansen, "Histological Evaluation of the Bone Response to Calcium Phosphate Cement Implanted in Cortical Bone," *Biomaterials*, **24**, 989-1000 (2003).

[19]B. Flautre, C. Maynou, J. Lemaitre, P. van Landuyt, and P. Hardouin, "Bone Colonization of β-TCP Granules Incorporated in Brushite Cements," *J. Biomed. Mat. Res.; Appl. Biomat.*, **63**, 413-417 (2002).

[20]D. Apelt, F. Theiss, A.O. El-Warrak, K. Zlinszky, R. Bettschart-Wolfisberger, M. Bohner, S. Matter, J.A. Auer, B. von Rechenberg, "In Vivo Behavior of Three Different Injectable Hydraulic Calcium Phosphate Cements," *Biomaterials*, **25**, 1439-1451 (2004).

[21]R. Tang, M. Hass, W. Wu, S. Gulde, and G.H. Nancollas, "Constant Composition Dissolution of Mixed Phases II. Selective Dissolution of Calcium Phosphates," *J. Coll. Int. Sci.*, **260**, 379-384 (2003).

[22]J.H. Hanks and R.E. Wallace, "Relation of Oxygen and Temperature in the Preservation of Tissues by Refrigeration," *Proc. Soc. Exp. Biol. Med.*, **71**, 196 (1949).

[23]T. Kokubo, "Surface chemistry of bioactive glass ceramics," *J. Noncryst. Solids*, **120**, 138-151 (1990).

[24]S.V. Dorozhkin, M. Schmitt, J.M. Bouler, and G. Daculsi, "Chemical Transformation of Some Biologically Relevant Calcium Phosphates in Aqueous Media during a Steam Sterilization," *Journal of Materials Science: Materials in Medicine*, **11**, 779-786 (2000).

[25]A.C. Tas, "Synthesis of Biomimetic Ca-Hydroxyapatite Powders at 37°C in Synthetic Body Fluids," *Biomaterials*, **21**, 1429-1438 (2000).

[26]L. Tortet, J.R. Gavarri, G. Nihoul, and A.J. Dianoux, "Study of Protonic Mobility in$CaHPO_4 \cdot 2H_2O$ (Brushite) and $CaHPO_4$ (Monetite) by Infrared Spectroscopy and Neutron Scattering," *Journal of Solid State Chemistry*, **132**, 6-16 (1997).

[27]V.S. Joshi and M.J. Joshi, "FTIR Spectroscopic, Thermal and Growth Morphological Studies of Calcium Hydrogen Phosphate Dihydrate Crystals," *Cryst. Res. Technol.*, **38**, 817-821 (2003).

[28]M. Trpkovska, B. Soptrajanov, and P. Malkov, "FTIR Reinvestigation of the Spectra of Synthetic Brushite and its Partially Deuterated Analogues," *J. Molec. Struc.*, **480-481**, 661-666 (1999).

[29]J. Xu, I.S. Butler, and D.F.R. Gilson, "FT-Raman and High-pressure Infrared Spectroscopic Studies of CaHPO$_4$·2H$_2$O and CaHPO$_4$," *Spectrochimica Acta*, **A55**, 2801-2809 (1999).

[30]J. H. Lee, D. H. Lee, H. S. Ryu, B. S. Chang, K. S. Hong, and C. K. Lee, "Porous Beta-Calcium Pyrophosphate as a Bone Graft Substitute in a Canine Bone Defect Model," *Key Engineering Materials*, **240-2**, 399-402 (2003).

[31]R. Tang, C.A. Orme, and G.H. Nancollas, "A New Understanding of Demineralization: The Dynamics of Brushite Dissolution," *Journal of Physical Chemistry B*, **107**, 10653-10657 (2003).

[32]S.R. Kim and S.J. Park, "Effect of Additives on the Hydrolysis of Dicalcium Phosphate Dihydrate"; pp. 201-207 in *Ceramic Powder Science III*. Edited by G.L. Messing, S.I. Hirano, and H. Hausner. American Ceramic Society, Westerville, Ohio, 1990.

[33]H. Monma and T. Kamiya, "Preparation of Hydroxyapatite by the Hydrolysis of Brushite," *Journal of Materials Science*, **22**, 4247-4250 (1987).

[34]T. I. Ivanova, O. V. Frank-Kamenetskaya, A. B. Koltsov, and V. L. Ugolkov, "Crystal Structure of Calcium-Deficient Carbonated Hydroxyapatite. Thermal Decomposition," *J. Sol. State Chem.*, **160**, 340-349 (2001).

[35]K. Flade, C. Lau, M. Mertig, and W. Pompe, "Osteocalcin-Controlled Dissolution-Reprecipitation of Calcium Phosphate under Biomimetic Conditions," *Chem. Mater.*, **13**, 3596-3602 (2001).

[36]D. Knaack, Malleable Implant Containing Solid Element that Resorbs or Fractures to Provide Access Channels. US Patent No: 6,599,516. July 29, 2003.

Author Index

Keyword Index